Physical Oceanography

Developments Since 1950

Physical Oceanography
Developments Since 1950

Edited by

Markus Jochum
National Center for Atmospheric Research
Boulder, Colorado, USA

and

Raghu Murtugudde
University of Maryland
College Park, Maryland, USA

Springer

Markus Jochum
National Center for Atmospheric Research
Oceanography Section
1850 Table Mesa Drive, Room 415
Boulder, CO 80307-3200
USA
markus@ucar.edu

Raghu Murtugudde
ESSIC/UMD
CSS Bldg/Room 2201
College Park, MD 20742
USA
ragu@essic.umd.edu

Cover photograph courtesy of Andrew Shepherd *NOAA/PMEL*.

Library of Congress Control Number: 2005935458

ISBN 10: 0-387-30261-1
ISBN 13: 978-0387-30261-4

Printed in the United States of America. (BS/DH)

9 8 7 6 5 4 3 2 1

springer.com

Truth is a pathless land

Krishnamurti

Contributors

Francis Bretherton received his Ph.D. in Applied Mathematics from the University of Cambridge in 1961 and is currently Professor Emeritus at the University of Wisconsin.

Kirk Bryan received his Ph.D. in Meteorology from the Massachusetts Institute of Technology in 1958 and is currently Senior Research Scholar at Princeton University.

Russ E. Davis received his Ph.D. in Chemical Engineering from Stanford University in 1967 and is currently Research Oceanographer at the Scripps Institution of Oceanography.

J. S. Godfrey received his Ph.D. in Physics from Yale University in 1968 and is now retired in Tasmania.

Michael J. McPhaden received his Ph.D. in Physical Oceanography from the Scripps Institution of Oceanography in 1980 and is currently Senior Research Scientist at NOAA's Pacific Marine Environmental Laboratory.

Dennis Wilson Moore received his Ph.D. in Applied Mathematics from Harvard University in 1968 and is currently Leader of the Ocean Climate Research Division at NOAA's Pacific Marine Environmental Laboratory.

Walter Munk received his Ph.D. in Oceanography from the Scripps Institution of Oceanography in 1947 and holds the Secretary of the Navy Chair in Oceanography.

Joseph Pedlosky received his Ph.D. in Meteorology from the Massachusetts Institute of Technology in 1963 and is currently Senior Scientist at the Woods Hole Oceanographic Institution.

S. George Philander received his Ph.D. in Applied Mathematics from Harvard University in 1969 and is currently Professor at Princeton University.

Joseph L. Reid received his M.Sc. in Physical Oceanography from the University of California (Los Angeles) in 1951 and is currently Professor Emeritus at the Scripps Institution of Oceanography.

Bruce A. Warren received his Ph.D. in Physical Oceanography from the Massachusetts Institute of Technology in 1962 and is currently Scientist Emeritus at the Woods Hole Oceanographic Institution.

Carl Wunsch received his Ph.D. in Geophysics from the Massachusetts Institute of Technology in 1966 and is currently a Professor there.

Klaus Wyrtki received his Ph.D. from the University of Kiel in 1950 and is currently Professor Emeritus at the University of Hawaii.

Foreword

Over the last five decades Physical Oceanography developed explosively from a state with only a few observations and theories to a mature science with global field programs, massive computer power, and a complex theoretical framework. The scientists who led this development are already or will soon be retired. This collection of essays documents some of the breakthroughs and also tries to capture the spirit of exploration and excitement that accompanied these developments.

The original motivation for the present book came from our desire to understand the current social and scientific framework in which we work as physical oceanographers. Brief reflection makes it obvious that this framework *must* have historical roots. However, discussions about these roots with senior scientists only made the picture more complex and confusing. We came to the conclusion that there is no simple story that explains the current state of affairs. The natural solution was to let senior scientists tell how they perceived the developments in the field, each from their own unique point of view. Thus, by surrendering editorial objectivity we arrived at a broader, more objective view. The approach is comparable to data acquisition: it is known that there are no perfect observations, so one makes many.

The goal then is to reduce biases by sampling as often as possible. However, to keep the book at a manageable size and still give the individual authors space enough to cover several decades, we were limited to 10 to 20 authors whose contributions should not exceed 20 pages. Thus, the book is by no means a complete history of physical oceanography; many important scientists and subdisciplines of the field are not accounted for. Still, we tried to provide a coherent yet varied author base that is geographically diverse and evenly distributed among the following (rather artificial) categories: modelling, observations and theory. We did not solicit contributions from close colleagues. Twenty senior scientists were asked to contribute, 13 did; 5 declined because they did not have sufficient time and 2 did not think the project to be a good idea in the first place. Unfortunately, there is a geographical bias in the latter two groups, rendering the authorship biased towards the United States.

From the beginning of the project every author knew the complete list of authors. The authors were encouraged to write personal views of the history of physical oceanography, the exception being Bruce Warren to whom we are indebted for the historical introduction. In line with the idea of abandoning objectivity, there was no formal review process. This makes the success of this book rely on the moral integrity of the authors, and we believe that each author made an effort to be subjective yet fair. The function of the editors was largely restricted to asking for clarifications in the submitted manuscripts.

Editing this book and the related discussions with many scientists proved exciting and illuminating. It illustrated that science is a social enterprise shaped by a community of individuals and the occasional element of chance. It made clear to us that progress relies on communication across the boundaries of disciplines and is made over decades, not months. Based on the following chapters it appears to us that two things stand out in the history of physical oceanography: large programs, which seem to have the power to create a community out of individuals, and the person of Henry Stommel, who provided inspiration to many.

Markus Jochum
Raghu Murtugudde
Fall, 2005

ACKNOWLEDGEMENTS

This book would not have been written without the initial encouragement of Nelson Hogg, David Kaiser, Joe Pedlosky, and Carl Wunsch. The editors are also grateful to Peter Gent and Bill Large as invaluable sources of historical information and general advice, and to Gustavo Goni for advice on publishing books. The editors were supported by the National Center for Atmospheric Research (M.J.) and the University of Maryland (R.M.).

Contents

1. Historical Introduction: Oceanography of the General Circulation to the Middle of the Twentieth Century 1
Bruce A. Warren

2. Reminiscences of MODE 15
Francis Bretherton

3. Modeling Ocean Circulation: 1960–1990, the Weather Bureau, and Princeton 29
Kirk Bryan

4. Contributions to Global Ocean Observations 45
Russ E. Davis

5. On Reconciling Oceanic and Atmospheric Estimates of Surface Heat Fluxes—and on Validating Greenhouse Models 67
J. S. Godfrey

6. El Niño and Ocean Observations: A Personal History 79
Michael J. McPhaden

7. Reflections of an Equatorial Oceanographer 101
Dennis Wilson Moore

8. Ocean Acoustic Tomography: From a Stormy Start to an Uncertain Future 119
Walter Munk

9. A History of Thermocline Theory 139
Joseph Pedlosky

10. Sextant to Satellite: The Education of a Land-Based Oceanographer 153
S. George Philander

11. Some Advances and Retreats in the Study of Ocean Circulation since 1935 165
Joseph L. Reid

12. Towards the World Ocean Circulation Experiment and a Bit of Aftermath 181
Carl Wunsch

13. Interview with Klaus Wyrtki, 25 February 1999 203
Hans von Storch, Jürgen Sündermann, and Lorenz Magaard

Index 239

Chapter 1

Historical Introduction

Oceanography of the General Circulation to the Middle of the Twentieth Century

BRUCE A. WARREN

INTRODUCTION

One observes what he can, devises concepts he thinks helpful, and uses what physics he knows and computational facility he has to try to describe and make sense of the ocean circulation. So, by the middle of the twentieth century, the oceanography of the general circulation consisted mainly of (1) global maps of prevailing surface currents; (2) sections and maps of water properties; (3) velocity and transport estimates made from density measurements, the thermal-wind relation, and conjectured integration constants, plus geopotential topographies relative to arbitrary pressure surfaces; (4) the

---

Bruce A. Warren, Scientist Emeritus, Woods Hole Oceanographic Institution.

descriptive apparatus of water masses and their origins; and (5) the beginnings of physics.

1. SURFACE CURRENTS

Knowledge of surface currents came from mariners' observations of displacement imparted to ships. In the ninth century, ca. 846, Ibn Khurradādhbih (1865, p. 293) cited sailors' information about the semiannual reversal of the zonal currents in the northern Indian Ocean; later Arab geographers repeated his report (Warren, 1987). However, they never had any sense of horizontal circulatory gyres, and conceived of this phenomenon as a sort of annual tide, with high and low water alternating between east and west. They may have been influenced in making this misinterpretation by their physics teacher, Aristotle (1931, Book II, Chap. 1), who believed that the natural direction of "flow" for water was downward, and therefore, at its lowest point, in the ocean, it would merely be "swinging to and fro."

Further known discoveries of major ocean currents did not occur until the European voyages of exploration. In 1498 Columbus (1933, p. 38) found strong westward flow along the north coast of Venezuela (the Caribbean Current?), and supposed it to be part of a general westward movement of ocean water following the heavens—a medieval and Renaissance idea perhaps elaborated from Aristotle's spheres rotating about a stationary earth. The Portuguese rounding southern Africa at the end of the fifteenth century encountered the Agulhas Current, and Ponce de Leon came upon the Gulf Stream in 1513. Reports by Portuguese pilots (Mota, 1972) show that by the second quarter of the sixteenth century they had learned (somehow) that the (South) Equatorial Current ran all the way from the Gulf of Guinea to the Antilles. While Japanese writers and mapmakers from the seventeenth century onward had identified segments of the Kuroshio as it passed through the island chains south of Japan, nineteenth-century European cartographers seem to have been the first to recognize it as a long, continuous stream (Kawai, 1998).

These were strong currents, met mostly close to shore, and so were detectible with primitive methods of navigation. To disclose the predominantly zonal currents in the open ocean far from land, however, required means to determine longitude well, and both the method of lunar distances and accurate, sturdy chronometers did not become available until the late eighteenth century. Even then the accuracy was not great, only about half a degree. [The accuracy of celestial determinations of latitude was of order ten minutes of arc (and highly variable) in the sixteenth century (various comparisons are given by Morison, 1971); it improved to somewhat better than one minute in good observing conditions by mid-twentieth century, according to a retired sea-captain friend of mine.] With just a few fixes usually taken per day (none at all under cloudy skies, such as hang over the Gulf Stream), and with dead-reckoned estimates of displacement through the water of uncertain reliability, the calculations of current velocities were rough. Nevertheless, the nineteenth century saw generalized maps of surface currents, of varying quality, for individual oceans and for the world ocean. [Peterson *et al.* (1996) have assembled an extensive collection of surface-current maps, from earliest times to the end of the nineteenth century.]

A. G. Findlay, for one, drew comprehensive maps for the Atlantic and Pacific Oceans (1853) and for the Indian Ocean (1866) that displayed the North and South Equatorial Currents, the transpacific Equatorial Countercurrent, the subtropical gyres, the major western-boundary currents, and the seasonally reversing circulation in the North Indian Ocean. Off Somalia, for the summer monsoon season, he (1866, p. 102) even sketched a "great whirl" of current, which, with initial capitals, was the name adopted for the feature by twentieth-century oceanographers when they rediscovered it.

However, he botched the Pacific subarctic gyre (small wonder, given the shipping conditions there), he understated the east–west asymmetry of the gyres, and in 1853 he sent the Agulhas Current entirely into the South Atlantic, despite Rennell's (1832, p. 98) earlier report that the greater part of it turned back into the Indian Ocean, south of Africa, occasioning "great eddies" there (perhaps including the Agulhas Rings that drift into the South Atlantic?). By 1866 he had come around to Rennell's view. (Never mind that Rennell himself, in Chap. I, Sect. 2, Part 12 of his book, had made the same mistake as Findlay in 1853, and that it was in Chap. II, Sect. 1 that he gave the correct information—without comment on his contradiction of what he had written 55 pages earlier. Perhaps if he had lived to complete the work, he would have straightened out his presentation.) The importance of commercial traffic for oceanographic observations in those years underlay Findlay's (1866, p. 97) otherwise startling remark, "The Agulhas Current . . . after the Gulf Stream, was the first that attracted the investigation of sufficient data."

Mariners became aware of the Antarctic Circumpolar Current in the early nineteenth century (G. Deacon, 1984, p. 86), but Findlay (1853, 1866)—and others—charted it as a region of predominantly northeastward flow. Maybe, as Peterson *et al.* (1996) suggested, the (few?) estimated surface velocities consisted as much of Ekman drifts due to the strong westerlies as to the deeper reaching geostrophic flow.

Maps of prevailing surface currents achieved their finest quality in Schott's (1943) construction. He scaled his current vectors by their magnitudes, provided maps in the tropics for both summer and winter, got the sense of the Pacific subarctic gyre right, included the Weddell gyre, and, correctly, made his Circumpolar Current set generally south of east, except for its northward course along the east coast of South America. And he got everything else more or less right, too.

The Equatorial Undercurrents barely break the sea surface, and so they escaped notice as ship drift. Even so, using drogues, Buchanan detected the undercurrent in the eastern Atlantic in 1886. Unfortunately, his discovery was largely overlooked until the Pacific Equatorial Undercurrent was revealed in 1952 [Montgomery and Stroup

(1962) told the whole story, with references and quotations], and so it had no influence on the development of oceanography.

2. WATER PROPERTIES

Direct measurement of subsurface currents did not become feasible in any significant way until the mid-twentieth century, so oceanographers studied what they could actually measure, namely, water properties—at first only temperature, but then salinity and oxygen concentration as well. Before the Second World War, high-quality temperature measurements were mostly made with mercury thermometers. Until the introduction in 1874 of reversing thermometers for determining both temperature *in situ* and depth of observation (formerly estimated from amount of line payed out), subsurface temperatures were measured with maximum–minimum thermometers at depth, or from water samples brought to the surface in insulated water bottles. Maximum-minimum thermometers could never detect temperature inversions, and so, for example, the *Challenger* Expedition (1873–1876), using only these during the first part of the voyage, missed the temperature maximum associated with the North Atlantic Deep Water in the western South Atlantic. On the other hand, the warm deep water in the Southern Ocean had already been found in the late eighteenth and early nineteenth century with insulated water bottles (G. Deacon, 1984, p. 23).

But Nansen, engaged with his *Fram* Expedition to the Arctic (1893–1896), realized that, to get accurate *in situ* temperatures from insulated water bottles, the temperature read on deck had to be increased by the amount of adiabatic cooling the sample had undergone in being raised from depth. Ekman (1905) therefore prepared tables and graphs for the addition. Helland-Hansen (1912) then reversed the calculation, *subtracting* Ekman's increments from *in situ* temperatures to obtain a quantity preserved during adiabatic displacements, which, adopting current meteorological terminology, he called the potential temperature.

The salinity determinations made on the *Challenger* were actually shipboard measurements of specific gravity at a standard temperature, and were not very reliable. Since the relative proportions of the major ions dissolved in seawater had been found during the course of the nineteenth century to be virtually constant around the ocean, oceanographers in the 1900s began titrating for chlorinity instead. At very best (not always achieved) the accuracy of titration salinities was ± 0.02. This was adequate for studying the Atlantic, but it is close to the level of actual salinity variation in the deep Pacific, which was thus obscured from view until conductivity methods of salinity measurement, an order of magnitude more accurate, became practical in the 1950s.

Hydrographic stations, comprising discrete, serial observations of these properties in the vertical, either to middepth or to the bottom, were occupied along lines across currents, basins, or entire oceans. The sections of property fields so obtained showed the shapes and sizes of the vertical variations, and the quasi-horizontal layering of these features. From such evidence Merz and Wüst (1922) inferred a layered scheme of meridional circulation in the Atlantic which was not challenged until the 1960s. [I sketched earlier the development of ideas about the deep ocean circulation (Warren, 1981), and so will not say much more about it here.] Notable atlases displaying color plates of property sections were compiled by Wüst and Defant (1936) and Wattenberg (1939) for the German *Meteor* Expedition in the South Atlantic (1925–1927), and by Fuglister (1960) for the I.G.Y. (1957–1958) cruises in the Atlantic; they have afforded several generations of oceanographers immediate, vivid impressions of how the Atlantic Ocean is structured.

As observations accumulated, it became practical to prepare maps of the horizontal distribution of properties at depth as well as at the surface. In the *Challenger* Reports Buchan (1895) included global maps of temperature and specific gravity (salinity) at the sea surface, and temperature maps at depths to 1500 fathoms. The surface maps look pretty good, but there are bullets in the subsurface ones, which suggest some bad data. However, the maps at 500 to 900 fathoms clearly depict the westward-pointing, high-temperature tongue of Mediterranean outflow water. A Mediterranean undercurrent in the Strait of Gibraltar had been conjectured for two centuries (M. Deacon, 1971, Chap. 7), but it was not actually observed [with drogues and with temperature and salinity (specific gravity) measurements] until 1870 (Carpenter and Jeffreys, 1871); so Buchan's maps may have been the first to show its subsequent spreading in the open Atlantic. It was later depicted with much better observational coverage, in salinity as well as temperature, in the excellent maps of Wüst and Defant (1936).

At this time Murray (1899) constructed the first global map of bottom temperature, apparently at the request of a geologist investigating the contrasting geographies of living conditions between the near-surface and bottom waters for organisms found as fossils in the sediments. Bottom-property maps can be misleading about flow fields, though, because on account of stratification their tonguelike patterns register bottom topography as much as currents; but physical oceanographers have continued plotting them anyway—perhaps out of habit.

Other types of maps were tried too, in attempts to depict property variations in surfaces or layers that better parallel the flow field. Wüst (1935) identified "core layers"—layers in which a vertical extremum, like an oxygen maximum, occurred—and mapped the extreme values of the properties observed in these layers over the Atlantic. Montgomery (1938) advocated plotting on surfaces of constant potential density instead, because to the extent that potential density is conserved in the subsurface circulation, flow and the mixing of water properties should take place along such surfaces. These have at least the advantage over core layers that there are infinitely many of them; but when core-layer extrema do not track isopycnals (as they frequently do not), interpretation of one or the other may be labored. Moreover, both types of map are often ambiguous about the relative degrees to which mean advection and lateral mixing shape the property variations. Nevertheless, they have been useful in tracing property features back to sea-surface locations where characteristics are imparted by air–sea exchange, or to sites of deep sinking.

These subsurface, ocean-scale maps were never synoptic maps (the data were not collected at one time), and rarely maps of mean distributions (except in the North Atlantic not many observations were repeated nearby and then averaged), but composite maps (all reliable observations made whenever in an ocean plotted together and contoured). The fact that coherent composite maps with smoothly contourable features could be prepared at all argued for great stability in the property fields. While oceanographers were aware of some mesoscale variability (current meanders, or wiggles in isopleths on sections, if nothing else), the revelation brought about by direct, sustained current measurements that the motion field was generally dominated by its fluctuations (even with tides and internal waves filtered out), and that the ocean was full of eddies, came as a great surprise in the 1960s.

Coverage with hydrographic stations was concentrated in the Atlantic before midcentury because the Indian Ocean was far from laboratories equipped for prolonged voyages, and the Pacific was too broad for pre-1960s research vessels to make transoceanic sections like those in the Atlantic. But neither remoteness nor climatic severity discouraged the men of the Discovery Investigations from their long exploration of the Southern Ocean in the 1930s (G. Deacon, 1937; 1984, pp. 63–73).

Two technical developments around the time of the Second World War promoted rapid mapping of the path of the Gulf Stream, by tracking its sharp, subsurface, cross-stream temperature gradient. With the bathythermograph, temperature soundings could be taken to depths of 150–250 m from a ship underway, and with the Loran-A shore-based radio-navigation system, the ship's position could be determined at any time within some 1000 km of the east coasts of the United States and Canada, to an accuracy of about 1 km. Following the Stream east of Cape Hatteras disclosed that it was not at all the broad diffuse flow indicated there by the ship-drift maps, but a narrow, swift current taking sinuous, changing paths, whose meanders occasionally closed and broke off as great current rings (e.g., Fuglister and Worthington, 1951). In addition to changing drastically the characterization of the instantaneous Gulf Stream, these novel observations influenced later conceptions of the Kuroshio (Kawai, 1972), and, especially, of the Agulhas Retroflection and formation of Agulhas Rings (Lutjeharms *et al.*, 1992).

3. GEOSTROPHY AND DYNAMIC CALCULATIONS

Geostrophy entered oceanography surreptitiously, cloaked in Bjerknes's circulation theorem. Even *The Oceans* (Sverdrup *et al.*, 1942) does not include the term, let alone "thermal wind." Mohn (1885) had given the formula connecting (geostrophic) velocity to the slopes of isobaric surfaces, but when Sandström and Helland-Hansen (1903) came to work out procedures for computing the (geostrophic) velocities and volume transports related to the baroclinic density field ("dynamic calculations"), they started with the circulation theorem. Although the word "geostrophic" would not be coined for another thirteen years (Gold, 1963), it is still puzzling why they took such a roundabout route when the geostrophic approximation falls so readily out of the momentum equations. [When he reworked some of the material from *The Oceans* for meteorologists, Sverdrup (1942, pp. 97–98) made that approximation explicitly, and identified the equations with those of the "geostrophic" wind in the atmosphere. This is the earliest use of the term in an oceanographic context that I have found—and it is borderline.]

Unlike meteorologists, before satellite altimetry oceanographers were never able to measure horizontal pressure gradients directly at even a single level with sufficient accuracy for geostrophic-velocity determinations. In effect, they integrated the thermal-wind equation, with some assumption instead about the velocity at some reference surface. A common practice, for example, was to assume zero velocity at 1000 or 2000 m, which were convenient depths for terminating hydrographic casts if one weren't interested in the deep water. Since current speeds generally increase sharply upward through the thermocline, one could calculate reasonable near-surface speeds on that basis, and it did not much matter where in the deep water one placed his "level of motion." On the other hand, for calculating the full volume transport of a current, and, especially, for treating deep currents at all, the choice of integration constants was (and is) a much more delicate matter.

Wüst's (1924) comparison of dynamically calculated and directly measured velocities in the Florida Current was justly celebrated for demonstrating the utility of the dynamic method. Given modern strictness about significance of results, it is perhaps worthwhile, however, to recall what Wüst actually did. He used Pillsbury's current measurements which, on the classic section III across the Straits of Florida (Fowey Rocks to Gun Cay, occupied in 1885 and 1886), consisted of values at six depths between the surface and 130 fathoms on five anchor-stations, and at 200 fathoms as well on a sixth. Each measurement took one-half hour but was, in fact, an instantaneous reading; many repetitions were made and averaged, so the total time per station ranged from 38 to 555 hours. Pillsbury plotted vertical curves of the averaged speeds at each station, extrapolated the curves to the bottom (maximum depth about 900 m), and found a line of zero velocity running a little above the bottom. Wüst thought this was consistent with a temperature inversion reported near the bottom (now known to be spurious), and did his dynamic calculations with Pillsbury's null-line as the reference. For local hydrographic data he found three temperature–salinity stations from 1914, and eight temperature-only stations from 1878 for which he used the 1914 temperature–salinity correlation to get density values; from these he developed four velocity–depth curves. While there was thus a good deal of flexibility in contouring both the "calculated" and "observed" velocity sections (illustrated with the observation points deleted as Fig. 184 in *The Oceans*), the agreement was impressive. Indeed, his estimated volume transport through Section III, $26 \times 10^6\,\mathrm{m^3\,s^{-1}}$, is only some 20% less than the present-day measurement. Nevertheless, as much as one admires Wüst's ingenuity in finding and combining meager material to make a useful calculation, one would probably evaluate his comparison with some reserve nowadays.

There were a few attempts to conjure up oceanwide zero-velocity surfaces as references for geostrophic calculations; none was convincing in the end. In the Atlantic Defant (1941) identified a finite, continuous layer at varying intermediate depth in which (in effect) the horizontal density gradient was zero (or relatively small), and he proposed that the horizontal velocity itself should be zero there. Even though his idea was purely intuitive, with only esthetic justification, his reference-surface map was influential for a time—perhaps for lack of anything more plausible. Hank Stommel once asked him how he came to think it up. Defant's reply, as Stommel understood him, was that it was a "kosmischer Schwank" (cosmic prank). Many years later Fritz Schott suggested to me that Defant may really have said "kosmischer Schwung" (cosmic spring, or swing), because Defant had been known to recommend that one contour sparse data points with a "kosmischer Schwung" in the hand.

Whatever he said, Wüst, in several publications (e.g., 1958), used Defant's reference surface to calculate velocities and volume transports on the *Meteor* sections in the South Atlantic. His numerical values are generally doubtful, but he did show indisputably the strong western-boundary currents in the middepth and bottom-water layers, and his overall velocities and transports for them are probably not off by more than 50%.

Clowes (1933) made a more rational choice of zero-velocity surface when he calculated the transport of the Circumpolar Current through Drake Passage. Since the tracer-property observations of the Discovery Investigations showed that the current reached to great depth, he assumed zero velocity at 3500 m, which is roughly the depth of the passage. His computed transport was 110×10^6 $m^3 s^{-1}$, just 15% less than the modern, directly measured value there.

Maps of the geopotential topography of isobaric surfaces (especially the sea surface) relative to that of some deeper surface, by exhibiting patterns and strengths of relative geostrophic flow, sharpened current maps obtained from ship drifts, provided snapshots of surface currents, and helped to delineate subsurface flow fields. Reid's (1961) composite map of surface geostrophic flow for the whole Pacific, for example, defined more surely than could Schott (1943) the subarctic gyre and the flow in the central South Pacific and in the Southern Ocean; and it disclosed a narrow, weak South Equatorial Countercurrent.

Quasi-synoptic maps of surface dynamic topography (apparently relative to 800 decibars), derived from repeated surveys off Japan that had begun in the 1930s, first documented the famous great meander of the Kuroshio that comes and goes south of Honshu at intervals of several years (Uda, 1951, 1964). This remarkable phenomenon seems not even yet to have been explained satisfactorily. The most practical application of dynamic calculations was the preparation for many years by the International Ice Patrol of springtime maps of surface dynamic topography (usually relative to 1000 decibars) offshore of Labrador and Newfoundland to help predict the possible drift of icebergs into shipping lanes. (These nicely hand-drawn maps, included in the annual reports of the Ice Patrol in the U.S. Coast Guard Bulletins, are a reproach to the lazy clumsiness of some computer-generated mapping these days.)

4. WATER MASSES

The most formal conceptual device for describing water-property distributions is the water mass. Helland-Hansen (1916) observed that when the temperatures and salinities measured at individual stations were plotted against each other, the resulting curves were similar over broad regions (and markedly different from those obtained in widely separated regions). This fact suggested the useful concept of the "water mass," defined by some appropriate segment of a regional T–S curve, much as an air mass was defined by a temperature–humidity curve.

Since subsurface temperature and salinity characteristics derive ultimately from the air–sea exchange at the sea surface, different segments of T–S curves correspond to different sites and mechanisms of water-mass "formation," and are so labeled: e.g., North Atlantic Deep Water or Antarctic Bottom Water as water masses formed through deep sinking in high latitudes. At thermocline levels the characteristics are imparted at lower, subtropical latitudes; Iselin (1939) emphasized that different portions of (vertical) T–S curves in the North Atlantic thermocline match closely the late-winter T–S conditions in different regions of the sea surface there, and he suggested that the thermocline curves were formed through lateral mixing from the sea surface along isopycnal surfaces. Because oxygen is consumed in the decay of sinking detritus, levels of oxygen concentration can give a qualitative notion of the local "age" of a water mass, i.e., how long since the water was "renewed" by exposure to the atmosphere. Oxygen concentration is sometimes used, in fact, as a subsidiary diagnostic in the definition of a water mass.

Water-mass analysis provides a summary description of the property fields; by highlighting prominent features and tracing them back to the locations where they were generated, it helps explain them; and their distribution (similar in concept to that of core layers) gives an impression of the patterns of prevailing flow and mixing responsible for their spreading.

Unfortunately, the concept is somewhat amorphous. Different writers have used different schemes of nomenclature, and when enthusiasts for fine distinctions proliferate names across the ocean (especially names reduced to letter sequences) with such vigor that the bewildered reader needs a glossary, the enterprise founders. And some writers have misconceived their water masses more as objective building blocks or primordial solutions than as designations for features in continuous property fields.

The outstanding achievement of water-mass analysis was Sverdrup's famous Chapter XV in *The Oceans* (Sverdrup *et al.*, 1942). He contrived enough names to cover the major features of the world ocean—but not so many as to befuddle a reader—and designed them to relate different ocean regions (e.g., "Central Water" of the South Atlantic or the North Pacific or the Indian Ocean). He combined this with a layout of the circulatory gyres, geostrophic estimates of transports of most of the major ocean currents, and a Merzian rendition of the vertical–meridional circulation. Chapter XV was a tour de force that has never been equaled (and probably could not be, given the vastly increased amount of observational and theoretical material that a mind needs to digest now). It was the treatise on the general ocean circulation for its day, and while much has been learned since then, it is still good reading.

In 1968 Jerome Namias remarked to me that air-mass analysis had been an intermediate stage in meteorology, that with the ability to run dynamical models of the atmospheric circulation, nobody talked air masses anymore. To some extent, watermassology is becoming superannuated too, as oceanographers have become capable of measuring routinely and abundantly much more than the traditional water properties, and as general circulation models reach toward usefulness. But it seems to me that, given the stable characteristic features of the water-property fields, at least the nomenclature will endure as useful, convenient shorthand.

5. PHYSICS

Aristotle went out with the seventeenth century, but discussions afterwards about circulation mechanics, despite Newtonian physics, tended to be qualitative and sterile, exemplified by the late-nineteenth-century squabble between Croll and Dr. Carpenter about the roles of wind and density forcing (M. Deacon, 1971, pp. 320–328). By the time of the Second World War circulation physics consisted essentially of just two ideas: geostrophy, which says nothing about what drives circulation, and the Ekman spiral, which showed that the direct effect of the wind stress was confined to a thin surface layer, and therefore left it still a mystery how the wind could drive the manifestly deep-penetrating circulatory gyres.

After the war Sverdrup (1947) demonstrated that the vertically integrated flow in the open ocean was driven by the *curl* of the wind stress, and Stommel (1948) discovered that the extreme asymmetry of the circulatory gyres was due to the meridional variation in the local vertical component of the earth's rotation vector. Carrier (Munk and Carrier, 1950) made the mathematics more tractable by casting the asymmetry as a western-boundary layer superposed on a lower-order interior flow field. [So far as I know, Charney (1955) was actually the first to use (though very unobtrusively) the now commonplace term, "boundary current."] Stommel (1957) then resolved the vertically integrated motion into near-surface Ekman transport and deep-reaching geostrophic flow; and he showed that what the wind-stress curl does is to force convergence of the Ekman transport, and that the resulting vertical velocity in turn drives a divergent geostrophic flow at depth. So at last the basic mechanism by which wind drives circulation was grasped, and it was the starting point for the subsequent flowering of steady-state and time-dependent circulation theory.

Meteorologists had assumed that the contribution of the ocean circulation to the global meridional energy transport was negligible compared to that of the atmosphere;

some oceanographers suspected otherwise. With a rough-and-ready calculation, Sverdrup (Sverdrup *et al.*, 1942, pp. 99–100) estimated that the northward energy transport in the North Atlantic subpolar gyre was about 10% of the global total at latitude 55°N. Jung (1952) argued that the overall oceanic share could be substantially greater on account of the vertical–meridional circulations; but such direct calculations were bedeviled by level-of-no-motion guesswork. Even with much better estimates of the ocean circulation and the sea-surface heat fluxes, the issue is still unsettled.

Until the introduction of high-speed computers in the 1960s, application of physics to circulation problems was severely restrained by the computing facilities available: namely, tables, graphs, slide rules, and desk calculators. One summer when I was an undergraduate employee at Woods Hole, Fritz Fuglister set me to do the velocity and transport calculations for his recent I.G.Y. sections across the South Atlantic. It took me a whole month to do a single section—a job requiring only a few seconds on a modern computer (apart from a few hours, perhaps, to format the data). Riley's (1951) failure to calculate satisfying property budgets for the Atlantic Ocean was surely due in part to the utter inability at the time to handle large masses of data quickly and flexibly.

Of course physical oceanographers found additional employment for computers. In the introduction to his edition of *Aubrey's Brief Lives*, Oliver Lawson-Dick (1949, p. xxix) wrote "...the adoption of the Arabic numerals at the time of the Renaissance opened up a whole new world of thought, which the men of the seventeenth century explored with voluptuous delight." Latter-day, computer-furnished oceanographers have been doing something similar.

BIBLIOGRAPHICAL NOTE

As part of the *Challenger* Reports, Murray (1895) wrote an excellent history of oceanic exploration, from antiquity to the end of the nineteenth century. Wüst (1964) summarized the major oceanographic expeditions from the voyage of the *Challenger* to 1960. To my mind, the best history of physical oceanography, for information, intellectual context, and critical judgment, is Margaret Deacon's (1971). Stommel's (1965) book, *The Gulf Stream*, although somewhat dated now, still displays the lively, powerful mind of a man vigorously engaged with his subject, and full of ideas about it.

REFERENCES

Aristotle. *Meteorologica*. Translated from the Greek into English by E. W. Webster, 1931. In: *The Works of Aristotle*, Vol. 3, W. D. Ross, ed., Oxford University Press, London.

Buchan, A., 1895. Report on oceanic circulation, based on the observations made on board H.M.S. Challenger, and other observations. In: *Report on the Scientific Results of the Voyage of H.M.S. Challenger during the years 1872–76. A Summary of the Scientific Results*, Second Part, Appendix (Physics and Chemistry, Part VIII), 38 pp.

Carpenter, W. B., and J. G. Jeffreys, 1871. Report on deep-sea researches carried on during the months of July, August, and September 1870, in H.M. Surveying-ship 'Porcupine'. *Proc. R. Soc. London* **19**, 146–221.

Charney, J. G., 1955. The Gulf Stream as an inertial boundary layer. *Proc. Nat. Acad. Sci. USA* **41**, 731–740.

Clowes, A. J., 1933. Influence of the Pacific on the circulation in the South-West Atlantic Ocean. *Nature* **131**, 189–191.

Columbus, C. *Select Documents Illustrating the Four Voyages of Columbus*. Spanish text edited and translated into English by C. Jane. Vol. II, The Third and Fourth Voyages, 1933. Hakluyt Society, Second Series, No. 70, London, 164 +lxxxix pp.

Deacon, G. E. R., 1937. The hydrology of the Southern Ocean. *Discovery Reports* **15**, 1–124.

Deacon, G., 1984. *The Antarctic circumpolar ocean*. Cambridge University Press, Cambridge, 180 + vii pp.

Deacon, M., 1971. *Scientists and the Sea 1650–1900. A Study of Marine Science*. Academic Press, London, 445 + xvi pp.

Defant, A., 1941. Quantitative Untersuchungen zur Statik und Dynamik des Atlantischen Ozeans. Die absolute Topographie des physikalischen Meeresniveaus und die Druckflächen sowie die Wasserbewegungen im Raum des Atlantischen Ozeans. In: *Wissenschaftliche Ergebnisse der Deutschen Atlantischen Expedition auf dem Forschungs- und Vermessungsschiff "Meteor" 1925–27* **6**, 2nd Part, 1, pp. 191–260.

Ekman, V. W., 1905. On the use of insulated water-bottles and reversing thermometers. *Publications de Circonstance, Conseil International pour l'Exploration de la Mer*, No. 23, 28 pp. + 2 pl.

Findlay, A. G., 1853. Oceanic currents and their connection with the proposed Central-America canals. *J. R. Geogr. Soc.* **23**, 217–242.

Findlay, A. G., 1866. *A Directory for the Navigation of the Indian Ocean*. Richard Holmes Laurie, London, 1062 + xxxvi pp.

Fuglister, F. C., 1960. Atlantic Ocean Atlas of Temperature and Salinity Profiles and Data from the International Geophysical Year of 1957–1958. *Woods Hole Oceanographic Institution Atlas Series*, **1**, 209 pp.

Fuglister, F. C., and L. V. Worthington, 1951. Some results of a multiple ship survey of the Gulf Stream. *Tellus* **3**, 1–14.

Gold, E., 1963. Origin of the word "geostrophic." *Bull. Am. Meteorol. Soc.* **44**, 249.

Helland-Hansen, B., 1912. The Ocean Waters. An Introduction to Physical Oceanography. 1. General Part (Methods). *Internationalen Revue der gesamten Hydrobiologie and Hydrographie* (Band III), Hydrographisches Supplement, Ser. 1, 84 pp.

Helland-Hansen, B., 1916. Nogen hydrografiske metoder. In: *Forhandlinger ved de 16 Skandinaviske Naturforskermøte*, 357–359.

Ibn Khurradādhbih, Abu'l Qāsim 'Ubayd Allāh 'Abd Allāh. Le Livre des Routes et des Provinces, par Ibn-Khordadbeh. Arabic text edited and translated into French by C. Barbier de Meynard, 1865, *Journal Asiatique*, Série 6, **5**, 5–127, 227–296, 446–532.

Iselin, C. O'D., 1939. The influence of vertical and lateral turbulence on the characteristics of the waters at mid-depth. *Trans. Am. Geophys. Union* **20**, 414–417.

Jung, G. H., 1952. Note on the meridional transport of energy by the oceans. *J. Mar. Res.* **11**, 139–146.

Kawai, H., 1972. Hydrography of the Kuroshio Extension. In: *Kuroshio: Its Physical Aspects*, H. Stommel and K. Yoshida, eds., University of Tokyo Press, Tokyo, 235–352.

Kawai, H., 1998. A brief history of recognition of the Kuroshio. *Prog. Oceanogr.* **41**, 505–578.

Lawson-Dick, O., 1949. *Aubrey's Brief Lives*. Mandarin, London, 341 + cx pp.

Lutjeharms, J. R. E., W. P. M. de Ruijter, and R. G. Peterson, 1992. Interbasin exchange and the Agulhas retroflection; the development of some oceanographic concepts. *Deep-Sea Res.* **39**, 1791–1807.

Merz, A., and G. Wüst, 1922. Die Atlantische Vertikalzirkulation. *Zeitschrift der Gesellschaft für Erdkunde zu Berlin*, Jahrgang 1922, 1–35.

Mohn, H., 1885. Die Strömungen des Europaischen Nordmeeres. *Petermanns Mittheilungen*, Erganzungsheft, Nr. 70.

Montgomery, R. B., 1938. Circulation in upper layers of southern North Atlantic deduced with use of isentropic analysis. *Pap. Phys. Oceanogr. Meteorol.* **6**, 2, 55 pp.

Montgomery, R. B., and E. D. Stroup, 1962. Equatorial Waters and Currents at 150° W in July–August 1952. *The Johns Hopkins Oceanographic Studies* 1. The Johns Hopkins Press, Baltimore, 68 pp.

Morison, S. E., 1971. *The European Discovery of America: The Northern Voyages A.D. 500–1600.* Oxford University Press, New York, 712 + xviii pp.

Mota, A. Teixera da, 1972. Atlantic winds and ocean currents in Portuguese nautical documents of the sixteenth century. *Proc. R. Soc. Edinburgh Sect. B* **73**, 59–67.

Munk, W. H., and G. F. Carrier, 1950. The wind-driven circulation in ocean basins of various shapes. *Tellus* **2**, 158–167.

Murray, J., 1895. Historical introduction. In: *Report on the Scientific Results of the Voyage of H.M.S. Challenger during the Years 1872–76. A Summary of the Scientific Results*, First Part, pp. 1–106E.

Murray, J., 1899. On the temperature of the floor of the ocean, and of the surface waters of the ocean. *Geogr. J. London* **14**, 34–51.

Peterson, R. G., L. Stramma, and G. Kortum, 1996. Early concepts and charts of ocean circulation. *Prog. Oceanogr.* **37**, 1–115.

Reid, J. L., 1961. On the geostrophic flow at the surface of the Pacific Ocean with respect to the 1,000–decibar surface. *Tellus* **13**, 489–502.

Rennell, J., 1832. *An Investigation of the Currents of the Atlantic Ocean, and of those which Prevail between the Indian Ocean and the Atlantic.* J.G.A.F. Rivington, London, 359 + viii pp. + 2 pl.

Riley, G. A., 1951. Oxygen, phosphate and nitrate in the Atlantic Ocean. *Bull. Bingham Oceanogr. Collect.* **13**, 1, 126 pp.

Sandström, J. W., and B. Helland-Hansen, 1903. Ueber die Berechnung von Meeresströmungen. *Report on Norwegian Fishery- and Marine-Investigations*, **2**, 4, 43 pp.

Schott, G., 1943. Nachtrag zu dem Aufsatz "Die Grundlagen einer Weltkarte der Meeresströmungen". *Ann. Hydrogr. Marit. Meterol.* **71**, 281–282 and Tafel 22.

Stommel, H., 1948. The westward intensification of wind-driven ocean currents. *Trans. Am. Geophys. Union* **29**, 202–206.

Stommel, H., 1957. A survey of ocean current theory. *Deep-Sea Res.* **4**, 149–184.

Stommel, H., 1965. *The Gulf Stream: A Physical and Dynamical Description*, 2nd edition. University of California Press, Berkeley, 248 + xiv pp.

Sverdrup, H. U., 1942. *Oceanography for Meteorologists*. Prentice–Hall, New York, 246 + xv pp.

Sverdrup, H. U., 1947. Wind-driven currents in a baroclinic ocean, with application to the equatorial currents of the eastern Pacific. *Proc. Nat. Acad. Sci. USA* **33**, 318–326.

Sverdrup, H. U., M. W. Johnson, and R. H. Fleming, 1942. *The Oceans: Their Physics, Chemistry, and General Biology*. Prentice–Hall, Englewood Cliffs, NJ, 1087 + x pp.

Uda, M., 1951. On the fluctuation of the main stream axis and its boundary line of Kuroshio. *J. Oceanogr. Soc. Japan* **6**, 181–189.

Uda, M., 1964. On the nature of the Kuroshio, its origin and meanders. In: *Studies on Oceanography*, K. Yoshida, ed., University of Tokyo Press, Tokyo, 89–107.

Warren, B. A., 1981. Deep circulation of the world ocean. In: *Evolution of Physical Oceanography, Scientific Surveys in Honor of Henry Stommel,* B. A. Warren and C. Wunsch, eds., MIT Press, Cambridge, MA, 6–41.

Warren, B. A., 1987. Ancient and medieval records of the monsoon winds and currents of the Indian Ocean. In: *Monsoons*, J. S. Fein and P. L. Stephens, eds., Wiley–Interscience, New York, 137–158.

Wattenberg, H., 1939. Atlas zu: Die Verteilung des Sauerstoffs im Atlantischen Ozean. In: *Wissenschaftliche Ergebnisse der Deutschen Atlantischen Expedition auf dem Forschungs- und Vermessungsschiff*

"Meteor" 1925–1927 **9**, Atlas, 72 pl. (*Atlas of the Dissolved Oxygen Content of the Atlantic Ocean*, 1992, Amerind, New Delhi, 72 pl.)

Wüst, G., 1924. Florida- und Antillenstrom: eine hydrodynamische Untersuchung. *Veröffentlichungen des Instituts für Meereskunde an der Universität Berlin, Neue Folge. A. Geographisch-naturwissenschaftliche Reihe*, 12, 48 pp.

Wüst, G., 1935. Schichtung und Zirkulation des Atlantischen Ozeans. Die Stratosphäre. In: *Wissenschaftliche Ergebnisse der Deutschen Atlantischen Expedition auf dem Forschungs- und Vermessungsschiff "Meteor" 1925–1927* **6**, 1st Part, 2, 180 pp. (*The Stratosphere of the Atlantic Ocean*, W. J. Emery, ed., 1978, Amerind, New Delhi, 112 pp.)

Wüst, G., 1958. Über Stromgeschwindigkeiten und Strommengen in der Atlantischen Tiefsee. *Geol. Rundsch.* **47**, 187–195.

Wüst, G., 1964. The major deep-sea expeditions and research vessels 1873–1960. *Prog. Oceanogr.* **2**, 1–52.

Wüst, G., and A. Defant, 1936. Atlas zur Schichtung und Zirkulation des Atlantischen Ozeans. Schnitte und Karten von Temperatur, Salzgehalt und Dichte. In: *Wissenschaftliche Ergebnisse der Deutschen Atlantischen Expedition auf dem Forschungs- und Vermessungsschiff "Meteor" 1925–1927* **6**, Atlas, 103 pl. (*Atlas of the Stratification and Circulation of the Atlantic Ocean*, 1993, Amerind, New Delhi, 103 pl.)

Chapter 2

Reminiscences of MODE

FRANCIS BRETHERTON

PURPOSE

The specific topic of this memoir is the Mid Ocean Dynamics Experiment (MODE), later renamed MODE-1. This was a multi-investigator collaborative research program funded by the National Science Foundation during the first half of the 1970s as part of the International Decade of Ocean Exploration (IDOE), with additional support from the Office of Naval Research and the National Atmospheric and Oceanic Agency (NOAA). A complete documentation is beyond the reach of this author, who was one of the participating scientists with no formal responsibility for the program as a whole. Instead, the goal is to convey a sense of the excitement of bringing some theoretical tools learned in another field to bear on a basic scientific problem, and to sketch anecdotally some of the highlights of the group interactions that ensued. An underlying theme is the gradual transformation of physical oceanography from a discipline dominated by independent investigators, each with their private research program linked only by publication of final conclusions, into a collaborative enterprise in which talents, resources, ideas, and observational data must be pooled in order to address the fundamental problems.

Francis Bretherton, Professor Emeritus, University of Wisconsin.

DISCLAIMER

This account is based upon personal and fragmentary memories of a brief period (1970–1973) of intensive involvement by an applied mathematician with no formal training in oceanography. Few written records remain that are readily available to me and my recall has become more selective with age. The perspective is also unavoidably colored by a subsequent 3 decades of exposure to issues of climate change and the interactions of humans with our global environment. This later experience of "big science" as a way of life, superimposed on huge changes in technology and instrumentation, makes it difficult to wind back the scientific and interpersonal context to the MODE era in an objective manner. However, it is just this difficulty that makes the attempt worthwhile.

BACKGROUND

Following a Ph.D. in Fluid Dynamics from the Department of Applied Mathematics and Theoretical Physics (DAMTP) at the University of Cambridge in England, I spent a postdoctoral year (1961–62) visiting C.C. Lin's group in the Mathematics Department at the Massachusetts Institute of Technology. Sitting in on a course of lectures by George Veronis on Ocean Circulation, I had my first introduction to concepts such as Ekman layers, the level of no motion, and the Stommel model of western boundary currents. It was far from my area of specialization (classical fluid dynamics at low Reynolds number), but planted a seed that grew way beyond expectations.

Later that year, responding to an opportunity created by George Batchelor at DAMTP, I returned to Cambridge with a commitment to research in Dynamical Meteorology, as part of a new group focusing on atmosphere and ocean dynamics. For the first year my colleague in this endeavor was Owen Phillips, who later wrote the textbook *Dynamics of the Upper Ocean*. However, for me the first priority had to be mastering my new subject area! Nevertheless, the contact with George Veronis was invaluable, and at his invitation two summers (1963, 1964) were spent at the Woods Hole Oceanographic Institution, participating in the Summer Program in Geophysical Fluid Dynamics (GFD) in Walsh Cottage. Supported over many years by the National Science Foundation, this GFD program was highly influential in educating and recruiting graduate students and junior faculty into a burgeoning field. I certainly count myself in this category. Interaction with John Booker, one of the students in the program, gave rise to the concept of *critical layer absorption* for internal gravity waves in a shear flow (Booker and Bretherton, 1967), which became a substantial early contribution to my new field of research. Although my primary concerns were in the atmosphere, the equations of motion for a stratified fluid on a rotating earth were the same for both media (and indeed even within the sun), and it seemed that, given adjustment of space and time scales and appropriate observational tools, many phenomena should translate between them. It also should be noted that the intellectual hothouse of the scientific program gained substantially from the relaxed physical environment of Woods Hole in summer and from the assistance provided by Mary Thayer to families such as mine who were visiting from overseas.

During the 1960s computers were gradually introduced into routine weather prediction, an experience that has many parallels in present-day oceanography. At first, the models were not very sophisticated, with one or two layers in the vertical, generally assuming quasi-geostrophic balance, and for every forecast they had to be reinitialized from a traditional, hand-drawn analysis. I recall visiting the Royal Meteorological Office in Bracknell, England (probably in 1964). By chance it was the first day that the new automated data system and analysis procedure was operational. Everyone was very proud of this achievement, so I was surprised to find the Chief Duty Forecaster at the bench redrawing the 12Z weather map by hand. He explained that the latest satellite cloud image showed that the cold front in the mid-Atlantic had been placed too far to the west. However, to communicate this information to the computer, he had to generate bogus ship reports in the required international data format, and feed them into the automated data system for the next analysis cycle. There were also concerns about the data from a ship that had reported its position to be in the middle of Greenland. For me, these were sobering reminders that the practical benefits of using computers to solve the equations of motion would be realized only after painstaking attention to the realities of observation systems, particularly the characteristics and incompatibilities of the information they do or do not provide. Forty years later that wheel has come full circle. Computing capabilities have become enormously more powerful. Following sustained developments efforts by the meteorological services of the world, a systematic framework has been developed for addressing issues such as quality control on the available data streams, the adequacy of sampling, and the blending of measurements of different types within a numerical model into dynamically coherent representation of the atmosphere that evolves continuously on a daily basis. Indeed, these automated operational systems have become so pervasive and complex that they are impenetrable by outsiders, and their representations are often treated as fact, with only lip service paid to the shortcomings of the underlying model and to the assumptions governing the data screening and quality control.

In 1969 I moved to the Johns Hopkins University in Baltimore, Maryland, in the Department of Earth and Planetary Sciences.

THE ROOTS OF MODE

Soon after my arrival in Baltimore, Henry Stommel described his ideas for a project to pursue the intriguing discovery made a decade earlier by John Swallow and Hal Worthington (1957).

At that time our only knowledge of the circulation in the deep ocean came from analyses of hydrographic soundings. These were water samples taken from bottles spaced along a nonconducting cable lowered by winch from research ships on the surface. Currents within and above the thermocline were inferred from assumed geostrophic balance between the gradient of pressure in a horizontal direction and the Coriolis force associated with motion relative to a rotating earth. The horizontal pressure gradient was based upon differences of temperature and salinity between sounding stations that were typically spaced hundreds of kilometers apart, converted to density using laboratory determinations of the equation of state for seawater, and hence to pressure (or equivalently dynamic height) using hydrostatic balance. This inferential process gives only vertical differences in horizontal velocity, unless in each water column there is a reference level at which the velocity itself can be independently determined. Unfortunately, the obvious candidate for reference level, the ocean surface, was precluded by wind drift, tidal fluctuations, and uncertainties of navigation, and no proven alternative methodology existed.

However, in the cold, dense, abyssal water below the main thermocline the temperature and salinity seemed remarkably uniform, with horizontal differences that in most places were measurable only on the scale of an ocean basin. It was also known that the carbon-14 age of water deep in much of the Atlantic Ocean was at least several hundred years, and in the Pacific it was over one thousand. These facts suggested that the abyssal water was relatively quiescent, so horizontal velocities from deep hydrographic soundings were typically reported relative to a "level of no motion" that was supposed to be representative of the entire layer below the thermocline. To account for the carbon-14 data, water that had sunk or mixed beneath the surface in the North Atlantic had to be moving southward through the ocean basin. However, the average velocity required was about 1 mm/s, which was well below the precision of measurement of the remainder of the sounding. Thus, it was widely supposed that the deep water in the open ocean was effectively quiescent and there was an identifiable "level of no motion" below the main thermocline.

Swallow (1955) had devised an ingenious float consisting of a scaffolding tube plugged at both ends, equipped with batteries and an acoustic transducer that pinged at intervals. Because the effective compressibility of such a device was less than that of sea water, it could be ballasted to sink to a predetermined depth where it became neutrally buoyant. Thereafter it would drift with the current at that depth while its position was determined by repeated triangulation relative to a research ship equipped with a directional acoustic antenna at the ocean surface. In consultation with Stommel, he and Worthington (1957) had tracked several such floats, in a region where they expected the water below the main thermocline to be effectively stationary. It was not! Instead they found sustained velocities of up to 10 cm/s in several different directions. It was clear that some unanticipated dynamical process was operating, but a much larger sample would be needed to determine what that was. Unfortunately, directional tracking from a dedicated ship was prohibitively cumbersome and expensive, so it was impractical to use this methodology on the scale required.

By the early 1970s several technological developments had progressed to the point at which reconsideration seemed worthwhile.

First was the internally recording moored current meter. Under the leadership of Nick Fofonoff and others at the Woods Hole Oceanographic Institution, battery-operated instruments were available that recorded internally the velocity (magnitude and direction) of the water relative to the instrument, sampled at whatever interval was appropriate for the application. For work in the deep ocean, a string of such instruments was attached to a cable suspended from a buoy a few hundred meters beneath the ocean surface and moored to ballast dropped onto the ocean floor. The subsurface mooring was to reduce motion induced by surface waves, and a buoyant polymer cable was used over most of its length to reduce the stresses on all parts of the system. Once deployed from a research vessel, a complex operation requiring an experienced crew, the mooring was left unattended for up to a year until it was revisited. Upon return, an acoustically activated release severed the connection to the ballast and the remainder of the mooring rose buoyantly to the surface. The development phase had been plagued by repeated, seemingly inexplicable, losses of the entire string. These were eventually resolved when a section of severed cable was recovered with a shark tooth embedded within it, leading to its replacement by steel in the upper part of the rig. There were clearly many potential uses for a capability like this. An early candidate was a spatial array of such strings of current meters, deployed specifically to map a limited area comprehensively in three-dimensional space and time.

A second development was the SOFAR float, due to Tom Rossby at the University of Rhode Island. He had adapted the concept of the Swallow float to tracking over basin scale distances, using pulses of low-frequency sound emitted in the SOFAR (SOund Fixing And Ranging) waveguide. This waveguide surrounds the depth of the minimum speed of propagation that is found throughout most of the world ocean. In the region of interest its center is at a depth of about 1000 m. The low frequency of the sound reduces absorption by molecular processes. Higher speeds above and below refract the waves back towards the central depth so that they spread in two dimensions rather than three. Both these effects contribute to potential transmission over very long distances. To achieve this transmission requires a powerful transmitter at a depth close to the axis of the waveguide, and a sensitive receiver comparably located. In their original configuration, Rossby's floats were packed with batteries to power the transducer, and used listening stations deployed by the U.S. Navy for other purposes. By differencing the arrival times of a pulse at several listening stations, the horizontal position could be located several times a day to within less than 1 km, a capability that exceeded that of ship navigation on the ocean surface above. With a lifetimc of months to a year or more, a fleet of such floats could potentially provide an effectively continuous record of the absolute velocity at a known reference level, and remove a key ambiguity in the hydrographic method for determining the circulation.

A third substantial improvement in capability was the STD (Salinity, Temperature, Depth). This instrument provided continuous measurements of these three variables on board ship as it was lowered on a cable from a winch as in a traditional hydrographic bottle sounding, except that a multi-core conducting cable is required.

It enabled an unprecedented view of the fine structure sampled during descent and ascent, and electronic processing greatly reduced the load on skilled technicians that was required for accurate salinity measurements. The instruments themselves were commercially marketed, but because of reservations about calibration they were supplemented during the MODE field programs by a limited number of bottle samples attached to the same cable. Although the dynamic heights needed for the scientific analysis were relatively insensitive to the fine structure, the onboard readout proved invaluable in adjusting operations from multiple ships in the light of the state of the ocean at the time.

A fourth enabling innovation was a programmatic decision made within the National Science Foundation to consider coordinated proposals from teams of Principal Investigators from several different institutions, and to review and support those selected as an integrated whole. This recognized that some of the fundamental problems in oceanography required a range of talents and technical resources that were beyond the reach of any one institution, and capabilities that could be sustained for longer than the duration of a traditional grant. The IDOE was launched as an experiment to foster new approaches and the administrative arrangements to support them.

However, most important of all were the leadership and insights of Henry Stommel himself. His gift for interpreting the data and complexities of ocean dynamics in terms of simple mathematical models was an ongoing inspiration to young scientists like me. Indeed a decade previously he and collaborators (e.g., Stommel and Arons, 1960) had developed a theory of the abyssal circulation in an ocean with a flat bottom that was in direct contradiction to the observations of Swallow and Worthington. Driven by a postulated widespread upwelling at the base of the main thermocline, a geostrophically balanced circulation was linked to a deep western boundary current. However, it too predicted speeds in the interior region of at most a few millimeters per second.

The project he had in mind was in concept very straightforward, to saturate a typical small area of open ocean with enough instrumentation to map the fields of velocity and density, and to follow their evolution for long enough to assess the dynamical balances that were controlling them. When he asked me to consider how I might possibly contribute to a such a project my first thought was that I was a meteorologist rather than an oceanographer, but my second thought was that the relevant equations of motion were almost identical to those used in weather prediction (indeed simpler because of the absence of cloud), and that perhaps approaches familiar in the atmosphere might be applied to the ocean as well.

PREPARATION

Stommel assembled a team that included practitioners of a wide variety of observational techniques as well as a group of theoreticians committed to elucidating the dynamics associated with the anticipated observational results. Besides the current meters, SOFAR floats, and STD/hydrographic capabilities described in the previous section, there were many other instruments that promised additional types of information but were unready or not suitable for sustained large-scale deployment. Notable was John Swallow himself with a version of his original float that was equipped with a transponder. Tracking still required a dedicated ship, but was simpler and worked over longer ranges. There were two varieties of vertically profiling float, the first sensing the electric field associated with motion across the lines of the Earth's magnetic field, and the second measuring distance to previously planted acoustic transponders on the ocean floor. Comparison of ascending and descending profiles exposes the tides and internal waves, and allows qualitative inferences about the lower frequency velocities that were of most interest to MODE. There were stationary pressure gauges deployed on the sea floor that, after the tidal signal had been filtered out, could provide estimates of the changes with time of the pressure differences between pairs, and hence map the geostrophic velocity near the bottom. Unfortunately, the depth of the sea floor at each location cannot be determined with sufficient accuracy to provide a steady-state value for the velocity. There was also an inverted echo sounder, placed on the bottom and recording the transit time for an acoustic pulse reflected back from the ocean surface. This transit time responds predominantly to the changing depth of warmer water as the thermocline moves up and down. Apart from issues of data recovery, which still required retrieval by a ship, it was a simple and inexpensive instrument. It was of particular interest to me because a similar approach had become common in the atmospheric context. There a simple, ground-based dish antenna and sound source were used to monitor the height of the top of the well-mixed boundary layer above a land surface, using backscattering from temperature fluctuations in the inversion layer just above.

I was a member of the Theoretical Panel, with a project to develop a numerical model of a 480 km $\times$ 480 km square domain of the ocean, above bottom topography that was, or could be, representative of the area proposed for the MODE field program. My essential assistant in this was Mike Karweit. We called it a meso-scale model (Bretherton and Karweit, 1975) because it could resolve phenomena on the scale of the Rossby radius of deformation, which in the region of interest was around 40 km. It solved the quasi-geostrophic equations for a stratified fluid on a beta-plane, with an adjustable coefficient of friction at the ocean floor and a pseudo-viscosity that imposed on each Fourier component a decay rate proportional to its wave number to the fourth power. What was novel was the embedding in the surrounding ocean. All other existing models simulated an entire ocean basin with rigid boundaries, including western boundary currents as well as interior regions, and had a flat or highly simplified representation of the ocean floor. Given the limited computing power available at the time it was very difficult to do justice to both the boundary and the details of interior regions simultaneously. The artifice we used was to assume that, over the square domain of interest, the velocity fields in each layer could be represented by a Fourier series in each horizontal direction. This implies that the velocity at each point on the bounding surface of the domain is supposed equal to that at a corresponding opposite point. Such an idealization was widely used in representations of homogeneous turbulence. The spatial mean velocity and potential vorticity at each level were treated as external parameters, conceptually consistent with the corresponding area average that might be achieved by a fully resolving, whole basin, model. In the quasi-geostrophic approximation, the dynamic height for each layer serves as a stream function for that layer, and is not strictly periodic but may have a component that increases linearly across the domain and is associated with the area averaged velocity. For simplicity, I will refer to area averages as *the mean flow*, and the remainder of the Fourier series as *eddies*.

Initially the model domain had 32 x 32 grid points in the horizontal and 6 layers in the vertical, giving it a horizontal resolution of 15 km. The layers were conceptually each of uniform, predetermined, potential density and more closely spaced near the surface, to match observed stratification in the Sargasso Sea. Beta effects were added to the potential vorticity in all layers, whereas bottom topography directly affected that in the lowest layer only. The Fortran code, consisting of a box of punched cards, was developed on the local IBM7094 in Baltimore. For significant production runs it had to be hand carried to the National Center for Atmospheric Research (NCAR) in Boulder, Colorado, which operated the most powerful supercomputer then available to the research community. The local debugging process was extremely tedious. There was no on-line editor and the only available output was a 132-character line printer producing reams of paper covered with discrete symbols. With some chagrin but no surprise, I discovered after we had been running the model for 6 months that the bottom topography had been inserted with the wrong sign, i.e., all the hills were represented in the model as hollows, and all the valleys as ridges! Nevertheless, it did enable some illuminating experiments relevant to MODE.

The most dramatic of these experiments was to impose a sustained mean flow U on an initial state where everything was at rest, but with a density stratification that was realistic. If the mean flow were toward the west, eddies appeared immediately but soon settled down to a steady flow along smoothed contours of the topography plus equivalent beta, keeping the hills to the right. The degree of smoothing increased with the magnitude of U. On the other hand, if U was reversed the initial eddies rapidly became chaotic and their energy continued to grow approximately linearly in time, apparently without limit. They were most intense in the lowest layer. Analysis showed that the former case corresponds to the situation familiar in topographic boundary currents, with the dynamic height varying across the current as an increasing function of potential vorticity. In the latter case, the topographic boundary current would have to keep high ground to the left (in the Northern Hemisphere) and the dynamic height would decrease with increasing potential vorticity. This configuration is unstable, and as water parcels move randomly off their original contour they leave a stationary residual perturbation in potential vorticity of the lowest layer. This residue implies a systematic pressure force on the topography. The westward reaction to this force opposes the mean flow and provides the energy source for the growing eddies.

In another experiment starting from a randomly chosen initial eddy field, the net horizontal pressure force instantaneously exerted on the bottom was calculated in the model, and a barotropic mean velocity was allowed to develop freely in response, preserving the total energy of the eddies and mean flow as well as the mean density gradients. The eddy motion was chaotic, though because of bottom friction and lateral viscosity the overall intensity decayed slowly. Characteristic statistical properties of the eddies were monitored during this decay, to be compared in due course with the results of the MODE field program. They showed larger space and time scales above the thermocline, a minimum energy in middepths, and some intensification of the smallest scales near the bottom.

Another aspect of the Baltimore project was to prepare for introducing observational data into the model, either as initial conditions or on an ongoing basis. Mindful of that visit to the Royal Meteorological Office in Bracknell, it was anticipated that this might not be straightforward. In particular a complete initial state for the model required fields of dynamic heights at all grid points at all levels. Given this input data, the geostrophic velocities and potential vorticity necessary to proceed one time step could be derived in the model by finite differences, and the dynamic heights at the next time by solving a form of Poisson's equation. However, all the observational data would come from instruments that were at discrete, irregularly spaced locations rather than model grid points, and the measured variables would be either density or velocity, which convert into vertical or horizontal *gradients* of dynamic height, rather than of dynamic height itself. Thus, a consistent, automated procedure was needed to interpolate or extrapolate onto the grid.

The method chosen assumes that the underlying physical variables are governed by a joint normal multivariate probability distribution, with means and covariances that are known *a priori*. Then, if a number of observational data are provided containing independent, linearly additive errors that are also normally distributed with known mean and variance, Bayes theorem permits in principle calculation of the *a posteriori* probabilities of both the physical fields and the measurement errors, including the most probable value for each physical variable at each grid point and the uncertainty surrounding this value. This most probable value is a *maximum likelihood estimate* for the physical variable concerned. The calculation involves solving a potentially large number of linear simultaneous equations, but appropriate algorithms for doing so numerically were available. A major benefit of this approach is that it also permits calculation of the *mean square error of the estimate*, even if no actual data are available but only the underlying means and covariances. In particular, it provided an objective, computerized method of mapping in an intellectually consistent manner a collection of observations of different types onto a regular grid suitable for further analysis or modeling activities, together with a measure of the probable sampling uncertainties inherent in this mapping.

When I discussed this concept with Russ Davis, who was also a member of the Theoretical Panel, he pointed out that, provided attention is restricted to interpolation coefficients that do not depend on individual data but only on their statistics and array geometry, then the assumption of normal distributions was no longer necessary. The framework could then best be described as "*linear estimation with least mean square error*." He also later found that the central concepts had been anticipated in a meteorological context by Gandin in Leningrad in 1963 and indeed by Gauss in the nineteenth century (Bretherton, Davis, and Fandry, 1976). I have since noted several other disciplines in which the technique has been independently rediscovered.

Meanwhile, the remainder of the MODE Program was also making progress. In the Scientific Council, there was consensus that the primary objective was to map for a 3-month period the density and horizontal velocity fields throughout the water column below the seasonal thermocline, in an area away from the Gulf Stream that included both smooth abyssal plain and moderately hilly topography. The site was tentatively identified. A secondary objective was to assess the capabilities of promising instrument systems that might contribute to more extensive measurements in the future. Array 1 of current meters and hydrographic soundings was deployed to obtain preliminary measurements of what might be expected. A Program Office staffed by Dennis Moore was opened in Cambridge, Massachusetts. The Executive Committee began to meet every 2 weeks at MIT, and I had a loud, if naive, voice in its deliberations, during which I began to appreciate the complexities of organizing and coordinating an enterprise on this scale. On my return home to Baltimore in the evening, my daughter used to sniff the tobacco smoke in my clothing and exclaim, "Daddy, I know where you've been!"

The summer of 1972 was a critical time for finalizing the plans for the Field Program a year later. The Theoretical Panel met for 3 weeks at NCAR in Colorado with intense and sometimes heated discussion. On the side Mike Karweit was busy in the computing center and our modeling effort made great progress. Meanwhile the Observationalists deliberated less formally on the East Coast, though there was substantial cross-fertilization by visitors in both directions and individuals with roots in both groups. As I remember it, the major conclusion of the Theoretical Panel was that the proposed array of moorings at the center and corners of a hexagon of side 100 km was too widely spaced to allow mapping of velocities throughout the containing circle to an accuracy of $\pm 20\%$. This conclusion derived from estimates of expected mean square error computed from the measurements made in Array 1 (also known as MODE 0), using the technique of objective mapping described above. Unsurprisingly, it was also controversial, because most oceangoing scientists were accustomed to making their own judgments about such matters, and here was a bunch of theoreticians telling them what to do. After the ensuing debate the actual design retained the 100-km circle but three moorings were inserted closer to the center. In the event, Murphy's Law took its bite and a large number of the current meters in the main array failed, substantially degrading the outcome. The major achievement was that, in the MODE-1 Atlas (1977) that became the definitive summary of the observational results, for every variable there were objective analyses of the observed values, accompanied by maps providing at least a relative measure of the mean square error of the estimate.

THE MODE-1 FIELD PROGRAM

In March 1973 deployment of the main array began, and it was fully operational for April, May, and June. It did not pass without notice that the site was near the center of the infamous Bermuda triangle, where legend told of ships disappearing without trace. However, an unforeseen hazard proved to be modern airline service to Bermuda itself, which was essential as a staging base for equipment and other exceptional supplies destined for the ships at sea. During the winter and spring, the MODE Project Office had made extensive use of air freight, but when the tourist season began the airlines could make more money from passengers, and refused to accept any freight but meat, fresh fruit and vegetables. Dennis Moore spent countless hours on the telephone cajoling, bullying, and devising alternative routings.

I spent 3 weeks at sea helping with hydrostations on board a comfortable ship operated by NOAA. I must admit that on at least one occasion on my watch the STD showed evidence of mud from the ocean floor, but the data were apparently unaffected. I also saw with my own eyes the surface mooring at MODE Center that supported the only continuous wind record for the experiment. It was the only mooring in the entire array that was visible at the surface. Twenty-four hours later it had vanished, apparently *stolen* by a large unidentified ship seen stationary in its vicinity! Rumored informal enquiries failed to accomplish the return of at least its invaluable data, so the only measurements available to compute the wind forcing on the ocean surface during the field program were the standard weather observations from the visiting research ships.

Meanwhile the SOFAR float tracks and preliminary hydrostation data were unfolding on a daily basis. In Baltimore we used these data to construct and distribute preliminary maps of the stream function at float level. This effort was rewarded when a message from Tom Rossby at sea reported that a float that he had launched a few days previously had ceased transmission and he did not know where it was. Since these floats were very expensive he wished to recover it. He could communicate with it from the ship at a higher acoustic frequency, but only from within 1 km or so. Could we tell him where it might be found? The available data were slender in the extreme but we committed to the very first weather forecast for the abyssal ocean. To my intense surprise, he went to the predicted spot and recovered the float!

For me the final memorable event of the field program was in early July in Bermuda, where my family was on vacation while I did duty in the control center at the Marine Biological Center. Everything was running smoothly and there was little work to be done. Then came the first hurricane of the year, which was exceptionally early. The ships were never in danger, but the eye came straight toward Bermuda.

As we closed down everything and took refuge, we thought of the Bermuda triangle. Fortunately there was little damage, though the airport was closed for several days.

POSTSCRIPT

Three months later, I received a surprise call that could not be ignored. Effective immediately, I was asked to lead NCAR, an institution with some 300 employees including 100 scientists. Its mission is to support the university-based research community in the atmospheric sciences (broadly defined), helping them tackle the important problems that require facilities or resources beyond those that individual departments are able to support. It meant that I had to put my personal research on hold, drop out of active participation in MODE, and focus on mastering in a hurry a completely new job and its environment. One of my early actions was to establish a small Ocean Modeling Section at NCAR, and recruit Bill Holland and Jim McWilliams to join it. It was a decision without any regrets.

At NCAR, I gradually became aware of the pressing need for greater scientific understanding of the functioning of the Earth as a complete system, not only the interactions of the ocean and atmospheric circulations in climate, but also the biogeochemical cycles and biological processes on land and sea, all of which are being affected to an ever-increasing degree by human activities. This is a gigantic agenda, but national and international programs have been established, and substantial progress has already been made. My experiences in MODE were an invaluable lesson on "Big Science" in operation. By sharing them with others I hope to encourage present-day successors to pool their talents, and join together to address the really important issues that are facing their disciplines.

In the summer of 1974 the Dynamics Group (1975) met to discuss the implications of the MODE-1 observations, and relevant data from other studies that had recently become available. To my regret, I was unable to participate or even to focus on the outcome. In general terms, the upshot of this and other discussions seems to have been a consensus that the MODE-1 Field Program had proved the concept of mapping velocity and density fields on the meso-scale, and had demonstrated convincingly the capabilities or potential of many important new instruments. The presence in the MODE area of seemingly random eddies throughout the water column with abyssal velocities of several centimeters per second or more and time scales of weeks to months was beyond question. However, due primarily to sampling uncertainties, the accuracies achieved were inadequate to draw final conclusions about the dynamical processes involved. Nevertheless, stimulated by MODE activities, telltales of qualitatively similar meso-scale eddies were being found in records of many other ocean observations throughout at least the North Atlantic and these were being analyzed in a new light. It seemed likely that eddy activity to some degree was almost ubiquitous, though there were major gradients in intensity associated with proximity to the Gulf Stream. In conclusion, there was an immediate successor to the MODE program with many of the same participants. It was called POLYMODE but I am not qualified to write on the subject.

Thirty years later, it is accepted that quasi-geostrophic meso-scale eddies are in most parts of the global ocean and they are regarded as essential aspects of the circulation. Associated variations in sea level are routinely detected by satellite altimetry. Their essential characteristics can be simulated successfully in a global model driven primarily by wind stress on the surface, provided the spatial resolution is high enough. A principal source of their energy is baroclinic instability in regions of large horizontal gradients in potential density, but they interact strongly with local bottom topography where that is significant. They are responsible for most isopycnal mixing and lateral transfers of potential vorticity and momentum.

REFERENCES

Booker, J. R., and F. P. Bretherton, 1967. The critical layer for internal gravity waves in a shear flow *J. Fluid Mech.* **27**, 513–519.

Bretherton, F. P., and M. Karweit, 1975. Mid-ocean mesoscale modeling. In: *Numerical Models of Ocean Circulation.* National Academy of Sciences, Washington, DC, 237–249.

Bretherton, F. P., R. E. Davis, and C. B. Fandry, 1976. A technique for objective analysis and design of oceanographic measurements applied to MODE-73. *Deep-Sea Res.* **23**, 559–582.

Robinson, A. R., 1975. Dynamics and the analysis of MODE-1. Report of the MODE-1 Dynamics Group to the MODE-1 Scientific Council, MODE Executive Office, unpublished.

Swallow, J. C., 1955. A neutral-buoyancy float for measuring deep currents. *Deep-Sea Res.* **3**, 93–104.

Swallow, J. C., and L. V. Worthington, 1957. Measurements of deep currents in the western North Atlantic. *Nature* **179**, 1183–1184.

The MODE-1 Atlas Group, 1977. *Atlas of the Mid-Ocean Dynamics Experiment (MODE-1).* Massachusetts Institute of Technology, Cambridge, MA.

Chapter 3

Modeling Ocean Circulation

1960–1990, the Weather Bureau, and Princeton

KIRK BRYAN

Computing and numerical modeling have become such an important part of almost every scientific field, including oceanography, that it must be nearly impossible for a young person coming into the field today to imagine the precomputer era. Then as now, most physical oceanographers were concerned with the exploration of the oceans through measurements. There also existed a small group of theorists, and a few remarkable individuals with a foot in both camps, like Harrold Sverdrup, Walter Munk, and Henry Stommel. After a long incubation period numerical models of ocean circulation have now evolved into a third element in oceanographic research, partially bridging the gap between theory and observations and forming links with related disciplines. The original impetus for developing ocean models came from outside the oceanographic research community. As far back as the 1920s L. F. Richardson (1922) had a vision of mathematical models of the Earth's climate that would include the

Kirk Bryan, Senior Research Scholar, Princeton University.

Figure 3.1. At the dawn of the computer era, Joseph "Joe Smag" Smagorinsky (1924–2005) had the courage, imagination, and entrepreneurial skills to begin building on Richardson's original vision of a quantitative model of a climate system at the U.S. Weather Bureau. Photo taken in 1983.

atmosphere, ocean, land surface, and cryosphere. Richardson recognized that many gaps existed in the basic understanding of the climate system essential for such an endeavor. As he put it , "We are forced to occupy a house which is still in the hands of the builders." Modelers are still in this uncomfortable position today. However, climate modelers and field workers studying basic processes are now recognizing the benefits of close collaboration to fill in the gaps. When Richardson wrote his book, *Weather Prediction by Numerical Process*, there were no computers powerful enough to carry out the calculations he envisioned. Even if they had been available, there was a lack of knowledge of the branch of mathematics required to use computers to model the circulation of the ocean and atmosphere. Research during World War II gave the impetus to the development of computers and pioneering work began on numerical weather forecasting led by John von Neumann and Jules Charney at the Institute for Advanced Study (IAS). Joseph Smagorinsky (Figure 3.1), who had been a staff member at the IAS, returned to the U.S. Weather Bureau with the ambitious goal of reviving Richardson's vision of a mathematical model of the entire climate system, using what had been learned in numerical weather forecasting as basic building blocks.

Ocean circulation models are critically dependent on computer technology. Computer resources available to ocean modelers are still not adequate to fully resolve the wide spectrum of time and space scales of the ocean circulation. It is not possible to say after more than four decades since work began that ocean models are mature. For reasons that will be made clear there is still much work to do. This brief account will only be concerned with the three decades between 1960 and 1990. I have associated these three decades with the birth, infancy, and adolescence of ocean general circulation models (OGCMs), recognizing that the maturity of ocean models is still some time off, perhaps in the middle decades of the twenty-first century.

THE 1960s: THE BIRTH OF OGCMs

Smagorinsky's group had already been in existence four years when I received an invitation in 1960 to leave Woods Hole and join his group at the Weather Bureau. From its inception "Smag's laboratory" got extraordinary support from an organization famous at that time for its parsimony. The chief of the Weather Bureau, Francis Reichelderfer, and his deputy, Harry Wexler, took a personal interest in what we were doing. Top-of-the-line computers were expensive, and our lab was privileged to have the best available, while many of the Bureau's personnel worked away in ancient offices in an advanced state of decay. In the first year I shared a crowded office with Syukuru Manabe (Figure 3.2) and Douglas Lilly. At lunch hour all the desks were pushed together to make a ping-pong table. It was a very stimulating environment, despite the drab "government issue" office buildings. Ideas bounced back and forth like the ping-pong balls at lunchtime. My task was to build an ocean model that could eventually be coupled to the atmospheric model under construction by other members of the group. I was totally ignorant of the numerical methods needed to make an ocean

Figure 3.2. Syukuru ("Suki") Manabe and the author at the Weather Bureau's General Circulation Laboratory (later the NOAA Geophysical Fluid Dynamics Laboratory in Princeton). A collaboration in building ocean–atmosphere models began in the 1960s and continued for nearly 30 years. Photo taken around 1965

model, or any model for that matter. Doug Lilly was an excellent mentor, introducing me to the literature, and his own notation, which he had developed for analyzing conservation properties of numerical models. Another research group with the same aim, but with much less in the way of resources, started up at about the same time at the University of California in Los Angeles, headed by Yale Mintz. There was a continual flow of ideas between the two groups. The numerical methods of Akia Arakawa (1966) at UCLA were key to the success of the first attempts to model ocean circulation at the Weather Bureau.

It is hard today to imagine how primitive even the most advanced computers of the early 1960s were. A large room full of equipment was far less capable than the simplest modern laptop. The two-dimensional flow of an idealized wind-driven ocean model appeared to be a tractable problem. Linear models had been developed by Stommel and Munk and there was a good deal of speculation about the role of inertial terms in the equations of motion, which could not be easily studied by analytic methods. From a modern perspective the two-dimensional wind-driven ocean model is only of academic interest, but the successful extension of the model into a nonlinear domain (Bryan, 1963) excited a lot of interest. The next step, an extension to three-dimensions, was certainly premature considering the limited computing resources available. We were well aware of the difficulty of resolving the long time-scales, and wide spectrum of spatial scales of the ocean circulation. The prudent strategy would have been to hold off for a few years. However, that was not the philosophy in the laboratory. Joe "Smag" was an incurable optimist, and an early disciple of "Moore's Law," which nearly 40 years ago projected that the number of transistors on a computer chip would double every 2 years. Global models of the ocean were needed to construct global models of air–sea interaction, and we were going to build them, regardless of the limitations of existing computers.

At about this time Michael Cox (Figure 3.3) joined Smag's group. He was first hired as a computer operator. He then became an accomplished programmer, and eventually a leading ocean modeler. Mike Cox had extraordinary technical ability along with a great intellectual curiosity. Quite a few hours of the week he would be in the little library of the laboratory, and would constantly bring to my attention new references for the work we were doing. Together, we developed a three-dimensional model based on the primitive equations and incorporating Akia Arakawa's ideas in the numerical design (Bryan, 1969). This model with later contributions from many people is the remote ancestor of many of the ocean circulation models used in ocean–climate and ocean–geochemistry applications today (Griffies, 2004). Since Mike Cox and I could not hope to resolve all the scales of the ocean circulation, we used the newly minted and fashionable ideas of geophysical fluid dynamics to formulate "oceanlike" problems that were more feasible, but still allowed us to move forward (Bryan and Cox, 1967, 1968a,b). At the end of the 1960s other groups began to develop ocean circulation models. In the former Soviet Union Artem Sarkisyan developed a geostrophic model for the diagnosis of ocean circulation. Gunter Fischer and Jim O'Brien were active at the National Center for Atmospheric

Figure 3.3. Michael Cox (1941–1989) joined the U.S. Weather Bureau General Circulation Laboratory shortly before the move to Princeton. Mike started out as a computer operator. With little formal training he developed an exceptional ability to design and analyze numerical experiments with ocean circulation models. Photo taken 1975.

Research (NCAR) in developing wind-driven ocean circulation models. At the Weather Bureau we had the advantage of working closely with experts in numerical weather prediction, and benefiting from their experience in designing models, which filtered out unwanted high-speed signals. For example, another complete three-dimensional ocean circulation model based on the primitive equations of motion was developed by Pat Crowley (1968) at the Lawrence Livermore Laboratory. The Crowley model had many advanced features, but it contained no filtering approximations to remove the high-speed external gravity waves. Since very short time steps were required to resolve the external gravity waves, this limitation made the model impractical to run on the primitive computers then available.

Since the main motivation for building an ocean model was climate, we attempted a simulation of air–sea interaction. Our first project was a coupled atmosphere–ocean model in very idealized geometry (Manabe and Bryan, 1969). This paper was the start of a research collaboration with Manabe that was to continue for nearly three decades. Our first calculation was intended to illustrate how anomalies in ocean surface temperature caused by upwelling and other factors could lead to geographic shifts of precipitation anomalies from the ocean to land.

One theme in ocean modeling became clear early on. The sponsorship of this expensive enterprise was not in main-line oceanographic laboratories, but in universities like UCLA, and organizations like NCAR and the Weather Bureau with a major focus on climate. This had positive and negative consequences. On the positive side it broadened the support for oceanographic research and fostered a "global" type of thinking. On the negative side, the separation of modelers and seagoing oceanographers in different facilities slowed the exchange of ideas. There are encouraging signs that today there is a better integration of modeling and measurements. This synthesis has largely been brought about by large-scale research programs like WOCE.

THE 1970s: THE INFANCY OF OGCMs

Smagorinsky's laboratory, rechristened the Geophysical Fluid Dynamics Laboratory (GFDL), moved to the Forrestal Campus of Princeton University in 1968. At that time no comparable facilities existed in Europe or Japan. The move gave the laboratory, in partnership with the university, a wonderful opportunity to host overseas visitors interested in this new field. The work of the laboratory benefited greatly from these extended visits. Particularly valuable were the early contributions of Adrian Gill (Figure 3.4), David Anderson, and Herbert Huppert from Cambridge University. The impact of later visitors Claes Rooth and Juergen Willebrand is discussed in the next section. Adrian provided advice, great encouragement, and a wide perspective. We

Figure 3.4. Adrian Gill, F.R.S., (1937–1986) a warm friend and an inspirational scientist, was a repeat visitor to Princeton in the 1970s. Photo taken 1980.

worked on a study of the Southern Ocean (Gill and Bryan, 1971) that provided some key insights on dynamical constraints imposed by the Drake Passage.

Perhaps the first ocean-only calculation to have an immediate impact on current oceanographic research was a study of the Indian Ocean (Cox, 1970; Bryan, 1991). During the Indian Ocean Program Cox used the newly developed Weather Bureau ocean model to simulate the monsoon-driven seasonal cycle of surface currents in the geometry of the Indian Ocean. The calculation was an extremely ambitious one for the primitive computers of the day. To allow for the long adjustment times of the ocean's main thermocline, the calculation was carried out in several stages in which the numerical grid was refined at each stage. Cox showed how the newly discovered, large-scale, equatorially trapped waves interacted with coastal upwelling along the coast of Africa. This study illustrated how a numerical simulation could be a powerful tool to link ocean theory and observation. In retrospect, the Cox study makes another important point. Synoptic eddies in the ocean, the equivalent of atmospheric weather, vary greatly in horizontal scale with respect to latitude. At the equator they are quite large, of the order of hundreds of kilometers. In higher latitudes differences in stratification and the local Coriolis force make synoptic scales quite small, of the order of 5–50 kilometers. With limited computer power ocean circulation models could be much more successful in simulating the synoptic scales of tropical oceans than synoptic scales at higher latitudes. This point is borne out by the relatively early success of numerical simulations of the El Niño phenomenon by Philander (1990), while models are only recently reproducing the correct path of the Gulf Stream and the Gulf Stream extension in the North Atlantic (Smith *et al.,* 2000).

By the early 1970s numerical models of the ocean circulation were recognized in some circles, at least, as an important new field of research. A conference was organized by the U.S. National Academy of Sciences (NAS, 1975) at the University of New Hampshire in October 1972. The conference was small, but included scientists from the former Soviet Union, Sweden, and Germany. Each invited talk was accompanied by an invited review, and the discussion was documented. The conference report published in 1975 is a remarkable time capsule of the state of ocean circulation modeling, and what scientists thought about the subject in that evolutionary period. Sarkisyan from the Soviet Union used diagnostic calculations based on geostrophic balance, which suggested the importance of bottom pressure torques, where strong currents flow over bottom topography. At the time this was a very new idea, since most existing theories of ocean circulation only dealt with wind forcing at the surface. However, the difficulties of making accurate calculations of this effect aroused skepticism. Even today, four decades later, bottom pressure torques and their role in guiding the path of western boundary currents are controversial. The first attempts to model the entire World Ocean were shown in papers by Cox (NAS,1975) and Takano *et al.* (NAS,1975). Cox presented a computation in which hydrographic data were inserted as an initial condition. The model was integrated forward for a few years, allowing a simulation of the World Ocean circulation to develop in response to surface winds and the pressure forces implied by data. Cox's model used a $2^\circ \times 2^\circ$ latitude–longitude grid, an amazingly high resolution for the limited computing resources available. North Atlantic simulations with essentially the same experimental design are now being carried out with $1/10^\circ \times 1/10^\circ$ latitude–longitude spacing (Smith *et al.*, 2000). This amounts to a factor of 20 in improved horizontal resolution. The modern calculations are able to capture a rich range of synoptic variability, but it is amazing how much Cox's prototype numerical experiment, 30 years earlier, captures the essential features of the large-scale circulation of the World Ocean.

The Takano *et al.* (NAS,1975) model was developed at UCLA by Yale Mintz's group. Takano's model was not actually ready at the time of the conference, but appears in the Academy Report. It was based on a somewhat idealized geometry of the World Ocean. It represents one of the earliest calculations of the World Ocean circulation with a forward model, in which the temperature and salinity fields are predicted from a resting initial state. The model contained some important improvements in numerics, which greatly simplified filtering the external waves in the original (Bryan, 1969) model.

In recent years all publicly funded, nonclassified projects must publish their software, making it available to the community. This was not always the case, and Joe Smagorinsky, who was very advanced in his thinking in other ways, thought that publication with a detailed description of methods was sufficient. The old assembly language codes that he was accustomed to would be of little use to another group, in any case. Bert Semtner (Figure 3.5), who was one of the first students in the Princeton program, went out to UCLA to join Yale Mintz's group after getting his degree. While doing pathbreaking modeling of mesoscale eddies with Yale Mintz (Semtner and Mintz, 1977) and in collaboration with Allan Robinson at Harvard (Robinson

Figure 3.5. Bert Semtner was one the first students at Princeton. Bert has been a pioneer in developing global models of the ocean circulation and the first to distribute the model to the research community(Semtner, 1974). His high- resolution global ocean circulation simulations played an important role in justifying the World Ocean Circulation Experiment (WOCE). Photo taken 1999.

et al., 1977), Bert combined the best features of the UCLA ocean model and the GFDL model and distributed the new code (Semtner, 1974) . In Princeton we adopted Semtner's code, which was clearly superior to the one we were using. A little later Joe Smagorinsky came to me about a request he had received for a copy of our code, and told me that it was against lab policy to share code. Naturally he was a little surprised when I told him that at present we were not using our own code, but the UCLA code! Once the precedent was set by Bert Semtner, code distribution continued through the years. Cox (1984) distributed regular code updates through e-mail, which was just coming into use. A great boon to modelers were the global climatological data sets (Levitus, 1982; Hellerman and Rosenstein,1983). After Cox's untimely death in 1989, Ron Pacanowski (Pacanowski *et al.*, 1991) took over the task of updating and distributing what became known as the GFDL Modular Ocean Model (MOM). A detailed description of the structure and physics of this type of model is given in a recent book by Griffies (2004).

The 1970s was the decade of MODE, a U.S. program to measure mesoscale, ocean eddies. This program spawned some very important ocean modeling by Peter Rhines, William Holland, and others. Models of ocean circulation on a planetary scale were not in fashion because critics pointed out that there was not enough data available to verify such a model, and a model which could not resolve mesoscale eddies would not be valid in any case. To counter the prevailing thinking, Larry Lewis and I carried out a calculation which we called "A Water Mass Model of the World Ocean" (Bryan and Lewis, 1979). We felt that the main features of the temperature and salinity of the World Ocean were reasonably well known, even if the details of the ocean circulation were not. The goal was to make a more detailed calculation of the evolution of the main thermocline than had been possible for Takano *et al.* (NAS,1975), resolving the long time scales of adjustment of the main thermocline. The great difficulty of the calculation was the centuries of "spin up" time required to get a balanced state of the ocean. We used the method of grid refinement pioneered by Cox's Indian Ocean study. The final solution was on a $2° \times 2°$ latitude longitude grid. This resolution was no better than that used by Cox (NAS,1975) in his earlier global model, but Cox's case was an essentially diagnostic calculation, using observed data as input. The World Ocean forward model did not have the resolution to give a good simulation of western boundary currents, but the Bryan and Lewis calculation did model the main features of the global temperature and salinity fields, and gave a remarkable simulation of the formation of Antarctic Intermediate Water in the Southern Hemisphere.

After this World Ocean calculation the way seemed clear to couple the ocean and atmospheric models together to make a realistic global climate model. Manabe and I had already made several different calculations in idealized geometries, which tested ocean–atmosphere coupling, and even the incorporation of sea ice. However, simulating the present global climate in a coupled model turned out to be much more difficult than we expected. Without the constraint of fixed upper boundary conditions, the ocean component of the coupled model did many unexpected things. The deep water circulation in the model seemed to bear no resemblance to what was expected.

Since these calculations took months to carry out with the computers we had available, this result was extremely frustrating. It was only much later that we realized we were tilting at windmills. From a modern perspective it was obvious that the models we were using did not have enough resolution to give a realistic simulation of climate in a coupled model, without the rigid specification of upper ocean boundary conditions of the Bryan and Lewis (1979) simulation.

THE 1980s: THE ADOLESCENCE OF OGCMs

We continued to have trouble getting our global coupled model to give realistic simulations. However, we were able to get interesting results for climate models in more idealized geometries. For the more constrained case of symmetry across the equator Manabe and I were able to calculate equilibrium climates over a wide range of atmospheric CO_2 levels. For colder climates the thermohaline circulation decreased in strength, but for warmer climates the thermohaline circulation maintained or even increased in strength in spite of a reduced meridional temperature gradient. The key factor was found to be the increase of the thermal expansion coefficient in the ocean with increasing temperature. Larger thermal expansion allowed north–south density gradients to be maintained in the warmer climates in spite of reduced meridional temperature gradients (Manabe and Bryan, 1985). Two other results in idealized geometry anticipated more recent studies of global warming with more general models. Mike Spelman and I (Bryan and Spelman, 1985) showed that global warming in a simple coupled climate model would lead to an eventual collapse of the thermohaline circulation. In a second calculation it was shown that if one hemisphere had a higher ratio of land compared to ocean, the response to greenhouse warming would be much greater in the hemisphere with more land. It will soon be possible to test these early predictions against observations.

Claes Rooth (Figure 3.6) spent a sabbatical year in Princeton in the 1980s. With his broad interests Claes interacted with almost everyone in the laboratory. His most important contribution at Princeton was to question the "uniqueness" of our ocean model solutions. This was something we had not considered. He suggested that the same upper boundary conditions on an ocean model might allow more than one type of ocean circulation, an idea going back to Stommel's (1961) seminal paper on a box model of the thermohaline circulation. Claes helped to inspire an oft-cited study by Frank Bryan (1986) using an ocean circulation model in a simple basin extending to both hemispheres. In spite of identical upper boundary conditions for a model ocean in the two hemispheres, the preferred numerical solution was asymmetric across the equator. This gave Manabe and I a clue to what was going wrong in our coupled ocean–atmosphere models. The ocean-only and the atmosphere-only components of the model were able to simulate observations, but the looser boundary conditions in the coupled model allowed other, totally unrealistic solutions. The resolution of the ocean models was still too crude to provide realistic heat balances at the ocean

Figure 3.6. Claes Rooth of the University of Miami was a visitor in Princeton in 1981. His visit inspired work on isopycnal mixing and multi-equilibrium solutions in ocean models. Photo taken 2005.

surface. Manabe *et al.* (1991) use a stopgap remedy called the "flux adjusted" coupled model. In this parametrization the flux of heat and fresh water at the air–sea interface consists of two parts: one part is fixed at the rate which is obtained by an ocean-only model with realistic boundary conditions, and a second part is variable, depending on the interactive sea surface temperature and air temperature in the atmosphere just above the surface. While this remedy reduces the generality of the coupled model, it allowed the coupled model to have a realistic response to perturbations such as the effect of changes in greenhouse gases. Years of struggle finally paid off. At the time of the first IPCC (International Panel on Climate Change) report GFDL was the only laboratory with a global coupled ocean–atmosphere model.

Claes Rooth and Juergen Willebrand also contributed to the very important idea for improving the lateral mixing parametrization in the ocean circulation model. It had been pointed out that lateral mixing in the real ocean is largely along isopycnal surfaces, while in our GFDL model parametrization mixing largely took place on horizontal surfaces. The result was a spurious mixing across tilted density surfaces, unsupported by field data. As far back as the National Academy meeting in New Hampshire in 1972 (NAS,1975) George Veronis pointed out that this spurious horizontal mixing across tilted isopycnal surfaces could reverse the predicted upwelling in the main thermocline in the GFDL ocean model. Following up on Rooth and Willebrand's ideas, and an earlier paper by Solomon (1971), Redi (1982) formulated a parametrization in which the mixing takes place along density surfaces in a Cartesian coordinate model. Now this idea is best known as one element in a more comprehensive lateral mixing parametrization by Gent and McWilliams (1990). Designing the correct numerical implementation of this parametrization proved to be much more difficult, and the details have only recently been worked out (Griffies *et al.*, 1998).

At Wally Broeker's suggestion we brought two of his students to Princeton, Jorge Sarmiento and Robbie Toggweiler. This allowed us to explore the application of our ocean circulation models to geochemistry. The new field rapidly became a flourishing enterprise. In spite of the limited resolution of the models used, the early simulations of Sarmiento (1983) and Toggweiler *et al.* (1989a,b) offered promising correspondence with geochemical data and the best estimates at the time of the expected ocean sequestration of anthropogenic carbon dioxide. In the early 1980s the Max Planck Institute in Hamburg also became interested in ocean geochemical modeling. We developed a friendly two-way rivalry in this area. Ocean geochemical modeling grew with the participation of many more laboratories both in the United States and abroad.

Early versions of the coupled ocean–atmosphere models contained a model of passive sea-ice for the polar areas, but we were under no illusions as to its completeness. In the 1970s Bill Hibler became a frequent visitor to GFDL and we collaborated in coupling the ocean circulation model with his model of ice dynamics (Hibler and Bryan, 1984). The early tests of the coupled ocean and active sea-ice were very encouraging. They illustrated in a quantitative way how ocean currents were responsible for the observed position of the sea boundary in the subpolar North Atlantic. Gradually active sea-ice models are becoming a standard component of climate models, but it is very difficult to realize their full potential because the ocean's synoptic scales are so extremely small in the Arctic.

The GFDL model is based on Cartesian coordinates. In the 1980s two other ocean circulation model architectures became widely accepted. In the "sigma" models the vertical coordinate is normalized by the total depth. Pioneering development on this type of model was carried out by Blumberg and Mellor (1987) and it has become a nearly standard model for coastal and near shore applications. Bleck and Boudra(1981) developed the first ocean circulation model based on hybrid vertical coordinates. A Lagrangian vertical coordinate is used for the main thermocline, based on isopycnal surfaces, while a Cartesian coordinate is used for the mixed layer. Both architectures have unique advantages. The "sigma" model has a much better treatment of complex bottom topography, but has the disadvantage of inherent errors in the calculation of horizontal pressure gradients over steep bottom slopes. Since mixing and advection in the main thermocline largely take place along rather than across isopycnal surfaces, isopycnal surfaces appear to be the most "natural" coordinate for an ocean circulation model. While an isopycnal model is in principle the most accurate architecture, the practical implementation is actually quite complicated due to the complex density equation of sea water. Two decades after these two new architectures were originally introduced, the full potential of either one has not been completely realized for simulating ocean circulation and climate. Much more effort is needed in this area.

In the 1980s the National Center for Atmospheric Research became a major laboratory for ocean circulation modeling. GFDL alumni Bill Holland and later Bert Semtner and Frank Bryan joined the NCAR staff. Peter Rhines and Francis Bretherton, while not directly involved in large-scale ocean modeling, had an important influence. Bill Holland and Frank Bryan led an effort to follow up early studies (Holland and Lin, 1975) of the North Atlantic with geostrophic models with more ambitious, "eddy permitting" calculations using the primitive equations in a version of Cox's model (Bryan and Holland, 1989). At about the same time, Bert Semtner with Bob Chervin (Semtner and Chervin, 1992) carried out similar "eddy permitting" calculations for the entire World Ocean in the most ambitious application of computers up to that time. The North Atlantic studies of Holland, and the World Ocean study of Semtner and Chervin set a pattern, which have been continued up to the present with higher and higher spatial resolution as more capable computers become available.

SUMMARY

This account only covers three decades, stopping at the nineties and is focused on Princeton and the Weather Bureau. By the 1990s ocean modeling became a truly international undertaking. Ocean and ocean–climate modeling was being carried out in so many places in this country and abroad that no one person can give a proper account. For recent developments in ocean circulation modeling, the reader is referred to a many-authored summary by Griffies *et al.* (2000).

As impressive as the recent progress in ocean modeling is, there might be questions in the atmospheric community as to why it has taken so long to reach this point. In the parlance of applied mathematics, the ocean circulation is a classic "stiff" system relative to the atmosphere. This means that the ocean has a wide range of spatial and temporal time-scales, one or two orders of magnitude greater than the atmosphere. Resolving all these space and time scales entails a great computational burden. Ocean modeling is as critically dependent on advances in computing technology as on the imagination and skill of modelers. As mighty as today's supercomputers seem to be, ocean modelers are still waiting for computers powerful enough to allow them to model the coupled ocean–atmosphere system in all its detail. Much of the present account dwells on the technical aspects of developing successive generations of numerical models and the spawning of new applications of the models to climate and geochemistry. Since the ocean is a complicated continuum, it is very hard to develop quantitative theories for its behavior and relationships to actual data. The power of models is to link theory and observation, but models cannot be thought of as a substitute for either theory or data. A very concrete scientific achievement of models is the ability to diagnose the thermohaline circulation from data, and to predict how it might change under different climate conditions. The ability of models to simulate the complicated tropical air–sea interaction of the El Niño phenomenon has led to many theories, which are still being tested (Chang *et al.*, 2005). It is through models that we can already see in a quantitative way that the World Ocean plays the dominate role in climate change and climate variability on longer time-scales.

REFERENCES

Arakawa, A., 1966. Computational design for long-term numerical integration of the equations of fluid flow: Two-dimensional incompressible flow. Part I. *J. Comput. Phys.*, **1**, 119–143.

Bleck, R., and D. Boudra, 1981. Initial testing of a numerical ocean circulation model using a hybrid (quasi-isopycnic) vertical coordinate. *J. Phys. Oceanogr.* **11**, 755–770.

Blumberg, A. F., and G. L. Mellor, 1987. A description of a three-dimensional coastal ocean circulation model. *Three-Dimensional Coastal Ocean Models*, Vol. 4, N. Heaps, ed., American Geophysical Union, Washington, DC, 208 pp.

Bryan, F., 1986. Maintenance and variability of the thermohaline circulation. Geophysical Fluid Dynamics Program Thesis, Princeton University.

Bryan, F., and W. R. Holland, 1989. A high resolution simulation of the wind and thermohaline-driven circulation in the North Atlantic Ocean. Proceedings of the 'Aha Huliko'a Hawaiian Workshop, January 17–20, 1989.

Bryan, K., 1963. A numerical investigation of a nonlinear model of a wind-driven ocean. *J. Atmos. Sci.* **20**, 594–606.

Bryan, K., 1969. A numerical method for the study of the circulation of the world ocean. *J. Comput. Phys.* **4**, 347–376, reprinted with corrections, *J. Comput. Phys.* 30th Anniversary Volume 1997.

Bryan, K., 1991. Michael Cox (1941–1989): His pioneering contributions to ocean circulation modeling. *J. Phys. Oceanogr.* **21**, 1259–1270.

Bryan, K., and M. D. Cox, 1967. A numerical investigation of the oceanic general circulation. *Tellus* **19**, 54–80.

Bryan, K., and M. D. Cox, 1968a. A nonlinear model of an ocean driven by wind and differential heating: Part I. Description of the three-dimensional velocity and density fields. *J. Atmos. Sci.* **25**, 945–967.

Bryan, K., and M. D. Cox, 1968b. A nonlinear model of an ocean driven by wind and differential heating: Part II. An analysis of the heat, vorticity, and energy balance. *J. Atmos. Sci.* **25**, 968–978.

Bryan, K., and L. J. Lewis, 1979. A water mass model of the world ocean. *J. Geophys. Res.* **84**, 2503–2517.

Bryan, K., and M. J. Spelman, 1985. The ocean's response to a CO_2-induced warming. *J. Geophys. Res.* 90, 11, 679–11,688.

Bryan, K., S. Manabe, and M. J. Spelman, 1988. Interhemispheric asymmetry in the transient response of a coupled ocean–atmosphere model to a CO_2 forcing. J. Phys. Oceanogr. **18**, 851–867.

Chang, P., T. Yamagata, P. Schopf, S. K. Behera, J. Carton, W. S. Kessler, G. Meyers, T. Qu, F. Schott, S. Sheytye, and S.-P. Xie, 2005. Climate fluctuations of tropical coupled system—the role of ocean dynamics. *J. Climate* (submitted).

Cox, M. D., 1970. A mathematical model of the Indian Ocean. *Deep-Sea Res.* **17**, 47–75.

Cox, M. D., 1984. A primitive equation, 3-dimensional model of the ocean. GFDL Ocean Group Technical Report No. 1, Geophysical Fluid Dynamics Laboratory, Princeton, NJ.

Cox, M. D., 1987. Isopycnal diffusion in a z-coordinate ocean model. *Ocean Modeling* **74**, 1–5.

Crowley, P., 1968. A global numerical ocean model: Part I. *J. Comp. of Phys.*, **3**, 111–147

Gent, P. R., and J. C. McWilliams, 1990. Isopycnal mixing in ocean circulation models. *J. Phys. Oceanogr.* **20**, 150–155.

Gill, A. E., and K. Bryan, 1971. Effects of geometry on the circulation of a three-dimensional southern hemisphere ocean model. *Deep-Sea Res.* **18**, 685–721.

Griffies, S. M., 2004. *Fundamentals of Ocean Climate Models.* Princeton University Press, Princeton, NJ, 518 pp.

Griffies, S. M., A. Gnanadesikan, R. C. Pacanowski, V. Larichev, J. K. Dukowicz, and R. D. Smith, 1998. Isoneutral diffusion in a z-coordinate ocean model. *J. Phys. Oceanogr.* **28**, 805–830.

Griffies, S. M., C. Boning, F. O. Bryan, E. P. Chassignet, R. Gerdes, H. Hasumi, A. Hirst, A.-M. Treguier, and D. Webb, 2000. Developments in ocean climate modeling. *Ocean Modelling* **2**, 123–192.

Hellerman, S., and M. Rosenstein, 1983. Normal monthly stress over the world ocean with error estimates. *J. Phys. Oceanogr.* **13**, 1093–1104.

Hibler, W. D., III, and K. Bryan, 1984. Ocean circulation: Its effect on seasonal sea-ice simulations, *Science* **224**, 489–492

Holland, W. R., and L. B. Lin, 1975. On the origin of mesoscale eddies and their contribution to the general circulation of the ocean. I. A preliminary numerical experiment. *J. Phys. Oceanogr.* **5**, 642–657

Holland, W. R., and J. C. McWilliams, 1987. Computer modeling in physical oceanography from the global circulation to turbulence. *Phys. Today* **40**, 51–57.

Levitus, S., 1982. Climatological atlas of the world ocean. NOAA Prof. Pap. 13, 173 pp. U.S. Government Printing Office, Washington, DC.

Manabe, S., and K. Bryan, 1969. Climate calculations with a combined ocean–atmosphere model. *J. Atmos. Sci.* **26**, 786–789.

Manabe, S., and K. Bryan, 1985. CO_2-induced change in a coupled ocean–atmosphere model and its paleoclimatic implications. *J. Geophys. Res.* **90**, 11,689–11,707.

Manabe, S., R. J. Stouffer, M. J. Spelman, and K. Bryan, 1991. Transient responses of a coupled ocean–atmosphere model to gradual changes of atmospheric CO_2, Part I: Annual mean response. *J. Climate* **4**, 785–818.

NAS, 1975. Proceedings of Symposium of Numerical Models of Ocean Circulation, Durham, NH. OSC/OAB, National Academy of Sciences.

Pacanowski, R. C., K. Dixon, and A. Rosati, 1991. The GFDL modular ocean model user guide, The GFDL Ocean Group Technical Report No. 2, Geophysical Fluid Dynamics Laboratory, Princeton, NJ.

Philander, S. G., 1990. *El Nino, La Nina, and the Southern Oscillation.* Academic Press, San Diego, 293 pp.

Redi, M. H., 1982. Oceanic isopycnal mixing by coordinate rotation. *J. Phys. Oceanogr.* **12**, 1154–1158.

Richardson, L. F., 1922. *Weather prediction by numerical process.* Cambridge University Press, London. Reprint, 1965 (With an introduction by Sydney Chapman). New York: Dover Publications.

Robinson, A. R., D. E. Harrison, Y. Mintz and A. J. Semtner, 1977. Eddies and the general circulation of an idealized oceanic gyre: A wind and thermally driven primitive equation numerical experiment.—2. Ocean eddies. *J. Phys. Oceanogr.* **7**, 182–207.

Sarmiento, J. L., 1983. A simulation of bomb tritium entry into the Atlantic Ocean. *J. Phys. Oceanogr* **13**, 1924–1939.

Semtner, A. J., Jr., 1974. An oceanic general circulation model with bottom topography. In: Numerical Simulation of Weather and Climate, Tech. Rep. No. 9, UCLA Department of Meteorology.

Semtner, A. J., and R. M. Chervin, 1992. Ocean general circulation from a global eddy-resolving ocean model. *J. Geophys. Res.* **97**, 5493–5550.

Semtner, A. J. and Y. Mintz, 1977. Numerical simulation of the Gulf Stream and mid-ocean eddies. *J. Phys. Oceanogr.* **7**, 208–230.

Smith, R. D., M. E. Maltrud, F. O. Bryan, and M. W. Hecht, 2000. Numerical simulation of the North Atlantic Ocean at 1/10 degree. *J. Phys. Oceanogr.* **30**, 1532–1561.

Solomon, H., 1971. On the representation of isentropic mixing in ocean circulation models. *J. Phys. Oceanogr.* **1**, 233–234

Stommel, H., 1961. Thermohaline convection with two stable regimes of flow. *Tellus* **13**, 224–230.

Toggweiler, J. R., K. Dixon, and K. Bryan, 1989a. Simulations of radiocarbon in a coarse-resolution world ocean model. 1. Steady state prebomb distribution. *J. Geophys. Res.* **94**, 8217–8242.

Toggweiler, J. R., K. Dixon, and K. Bryan, 1989b. Simulations of radiocarbon in a coarse-resolution world ocean model. 2. Distributions of bomb-produced carbon-14. *J. Geophys. Res.* **94**, 8243–8264.

Chapter 4

Contributions to Global Ocean Observations

RUSS E. DAVIS

My father, an attorney, often commented on the dynamics of legal contests: "More good stories are ruined by an eyewitness...." Over the last 40 years of oceanography to which I have been witness, a similar principle has held. Theory has explained why, and models have simulated, but most of the steps in knowledge have been closely tied to observations. In part, as Walter Munk has often said, this is because "Every time we look at the ocean in a different way we learn something new." Today, those interested in the large-scale structure and variability of the world's oceans are blessed with an amazing array of observations with which to work. Satellites provide the most voluminous ocean data with global coverage and have supported some of the most significant increments in understanding the ocean; the first SST images made vivid the different patterns of turbulence in different regions; today altimetry and scatterometry describe evolving dynamically important fields. Over the last half-century a complementary global *in situ* observing system has been developed and I have been fortunate to observe some of that development. I would like to tell what I know of its story.

Russ E. Davis, Research Oceanographer, Scripps Institution of Oceanography.

The capabilities of the emerging global ocean observing system have been widely described (see Koblinsky and Smith, 2001) and results from the more mature elements have already had important impacts on ocean science. Baker (1981) gives a wonderfully comprehensive, if dated, review of ocean observing instruments while Gould (2005) tells the story of subsurface floats from start to the present. McPhaden *et al.* (1998) describes the development of the tropical Pacific's Tropical Ocean Global Atmosphere (TOGA) observing system and McPhaden discusses it in this volume. To complement these, I focus on some of the ideas, people, and institutions that I think made progress possible. There are three themes: innovation, collaboration and teams, and luck.

THE PIONEERS

My introduction to oceanography in 1965 was great luck. A summer-program poster led me from engineering school to a laboratory in the Woods Hole Oceanographic Institution (WHOI) Smith Building working with Stuart Turner on how internal waves break. Fortune made the breaking process (through unstable nonlinear resonant interactions: Davis and Acrivos, 1967a) fascinating and understandable. More fortune led a fellow graduate student at Stanford to knock a screwdriver into a stratified wave tank, generating strange little eddies that were eventually identified as deep-water solitary internal waves (Davis and Acrivos, 1967b).

As I was being introduced to oceanography, development of today's ocean observing system was already in progress. In 1955 John Swallow had introduced the neutrally buoyant subsurface float to measure middepth currents. Until then, surface buoys tracked from ships and current meters lowered from anchored ships could measure only the strongest shallow flows, not the deep currents that were believed to be very weak. Swallow realized that aluminum pressure cases could be built to have density and compressibility that would allow them to equilibrate at a level of neutral buoyancy. If they could be tracked for days, very accurate measurements of deep currents would be possible. Swallow, a tall man of quiet manner and gentle humor working at Britain's National Institute of Oceanography, constructed the first floats from scaffolding tubing using caustic soda to thin the walls and attached surplus Royal Navy 10-kHz sound sources so that submerged floats could be located at ranges of a few kilometers from an accompanying ship using direction finding. Within 6 months Swallow completed design, construction, and field-testing of the first floats (Swallow, 1955) and two remarkable discoveries followed quickly.

Independently, Henry Stommel (1955) advocated determining deep currents by tracking neutrally buoyant floats using long-range acoustics in the SOFAR channel. In 1955 the float advocates met and agreed to test Stommel's (1957) idea, based on geostrophic shear and dynamical theory, of a deep equatorward countercurrent under the Gulf Stream. In an archetype of observational hypothesis testing, Swallow and Worthington (1957) confirmed the existence of this current, with speeds in excess of

Figure 4.1. John Swallow preparing an early acoustically tracked float for deployment. (Courtesy of John Gould.)

10 cm/s. In 1960 Swallow and James Crease examined another aspect of Stommel's proposition for deep circulation—the presumed broad poleward return flow for the western boundary undercurrent. Their study was a stunning example of influential discovery. The ketch *Aries* was used to track several long-lived floats deployed at 2 and 4 km depth 200 nm west of Bermuda. Expecting weak flow of O(1 cm/s), the *Aries* was scheduled over a period of 14 months to track floats at 2-day intervals with frequent several-day breaks to return to Bermuda. We now know that this area is the site of vigorous mesoscale eddies that caused the floats to move at O(10 cm/s) and, consequently, the experimental plan to be changed. The *Aries* results (Crease, 1962; Swallow, 1971) showed the energy of unexpected motions, with scales of around 70 km, and made it clear that the ocean was energetically turbulent. This observation had a profound effect on thinking about ocean dynamics and led to a vigorous research effort to understand the role of eddies in ocean circulation.

As Swallow floats were revolutionizing ideas of ocean circulation, Bill Richardson of WHOI was setting out on an even more challenging technical development: instrumented deep-sea moorings. Richardson *et al.* (1963) noted that (a) describing the currents that Swallow had shown to be highly variable would require long time series, (b) this was not practical by following floats with ships, and (c) instruments supported in the line below moored buoys could, in principle, obtain the needed long time series without the confusion between temporal and spatial variability found in float data. Richardson, who was creative and loved boats, airplanes, and adventure, set the audacious goal of using moorings to explore the circulation along a 1000-km line between Massachusetts and Bermuda. This would require moorings (i.e., anchor, mooring line, and flotation) strong enough to resist currents and instruments that could provide accurate measurements from these moorings. This remarkable effort, which lasted over a decade, was a fundamentally experimental war against Murphy's Law.

The Buoy Project began with the most difficult mooring type using a surface buoy subject to high loads from strong surface currents and surface waves. The surface float, a 2.9-m-diameter toroid with 3-ton displacement, had a rigid bridle connecting to the mooring line and a tower holding anemometer, light, and radio beacon. It was chosen for stability to overturn, simplicity of construction, strength, and ease of handling. The initial mooring line was polypropylene rope the buoyancy of which reduced peak line tension and surface-buoy size. Because loading caused the polypropylene to slowly creep and the conventional-lay rope to untwist (and in-line instruments to rotate rapidly), the line was switched to plaited nylon line with increased stretchiness, reducing wave loading and allowing the line's scope to increase with loading. Fastenings corroded at unpredictable rates that were apparently related to stress. Fish bite was identified as a cause of mooring failure so the upper mooring line was switched to torque-balanced steel wire. Anchors were hybrids of weight to resist mooring-line tension, chain to keep the mooring line off the bottom, and ground tackle to resist sliding over the bottom. Deployment was as follows. The surface buoy is deployed first, then the string of line and instruments is assembled as the mooring streamed behind the ship, and finally the anchor is dropped overboard, causing the surface buoy to race across the surface until it was over the anchor, sometimes moving 3 knots with a rooster tail behind it. Edward Brainard (1967) captured the challenge of Richardson's operations:

> My first impressions . . . were highlighted by the awesome feeling of great depths, 15,000–18,000 ft., tremendous Gulf Stream currents 4.5 knots or more, and the great loneliness of little 8 ft. diameter toroids sitting out in the vast cruel sea hundreds of miles from home.

With this perspective, one sees that although the average mooring life was barely 3 weeks, the early WHOI buoy work was a remarkable success. Indeed many of the components and techniques used in moorings today are evolutions from those developed by the Buoy Program.

Richardson left WHOI in 1963 and Buoy Group leadership passed to Nick Fofonoff and Ferris Webster. Fofonoff, who had already significantly advanced the theory of nonlinear ocean circulation, was an inspired choice who led the Buoy Project for more than a decade. During most of this time Bob Heinmiller managed the group as it methodically converted moorings from engineering challenge to measurement system; Heinmiller and Fofonoff (1995) describe this evolution. Initial surface mooring recovery involved a weak link that was broken by tensioning the mooring.

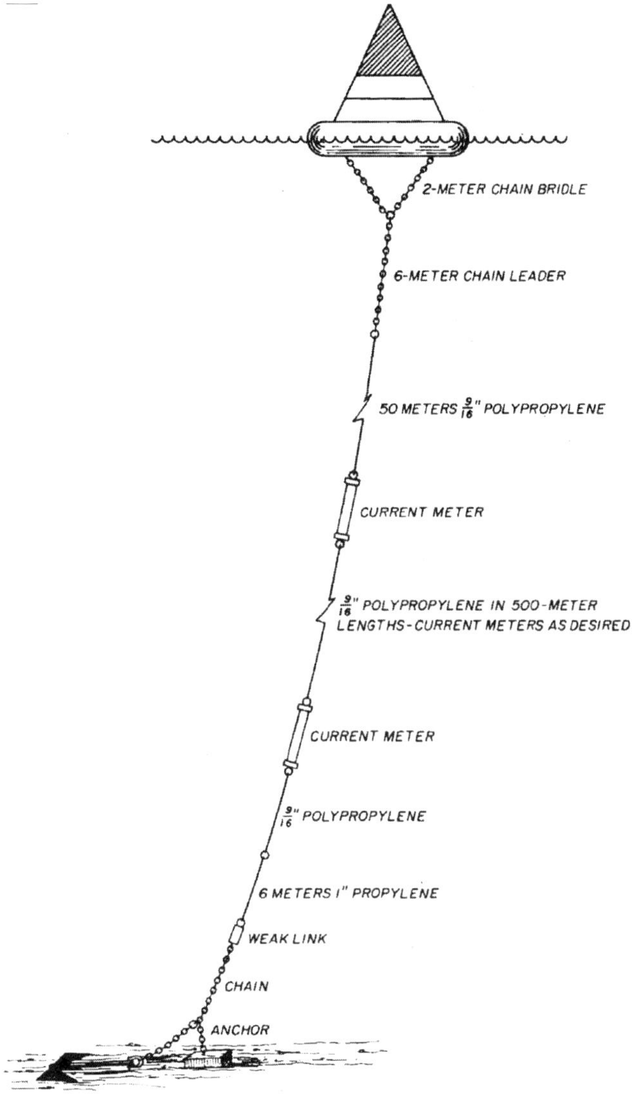

Figure 4.2. Schamatic of an early WHOI Buoy Group surface mooring from Richardson et al. (1963).

In conjunction with two manufacturers, the acoustically triggered anchor release was developed, making possible safer surface-mooring recovery and enabling subsurface moorings, which avoid wave forcing and its impacts on instruments. The original Richardson current meter used a Savonius-rotor speed sensor, compass and vane for direction, and, because low-power electronics were not available, an ingenious film recorder. Wave-induced mooring motion caused the apparent current direction to oscillate rapidly. This led to development of the Vector Averaging Current Meter (VACM) that used electronics to frequently sample speed and direction, convert them to vector components, average them, and record results on magnetic tape.

For many years the WHOI Buoy Group was the leader in both technical and scientific areas. Other groups were formed around the country using methods and equipment developed by the Buoy Group and employing personnel trained there. At the same time, Buoy Group work motivated an interest in observations and technical developments that made new measurements possible. It was an exciting center of observational oceanography.

During the 1960s, Swallow-float technology gradually improved but short tracking range made ships a continued necessity. In 1968 Tom Rossby and Doug Webb deployed two floats implementing Stommel's idea of long-range float tracking using sound propagation in the SOFAR channel (the sound-speed-minimum waveguide found near 1 km depth in tropical and subtropical oceans). At low frequencies sound absorption decreases so that tracking at ranges of 1000–2000 km becomes feasible, but the size of efficient low-frequency sound sources and batteries for extended duration led to large floats and difficult-to-build sound projectors. In the first test (Rossby and Webb, 1970) the floats projecting at 500 Hz were heard by a military tracking array at 1000-km ranges. Although these first floats failed after only a few days, a year later a 380-Hz SOFAR float was tracked for 4 months (Rossby and Webb, 1971). With tracking accuracies of O(3 km) limited by knowledge of the field of sound speed and much better day-to-day precision, this development made much of the world's ocean accessible to long-range, long-term direct velocity observation.

MODE—THE FIRST OCEAN OBSERVING SYSTEM

Prior to the 1970s, technology limited observational oceanography to individual surveys, single time series, or spatial sampling with very limited duration. The development of long-lasting floats and moorings made possible the Mid-Ocean Dynamics Experiment (MODE), which was the first large-scale observing effort that combined diverse observing tools into a designed system. The scientific focus (The MODE Group, 1978) was the mesoscale variability made so clear in the *Aries* observations that had subsequently been described by various other observations. The approach, led by Alan Robinson, Carl Wunsch, Francis Bretherton, and Henry Stommel, was methodical, using a sequence of focused efforts. MODE-0 used an irregular array to explore (a) sites for the MODE-1 array, (b) the coherence scales of mesoscale variability, and (c) spatial variation of eddy energy, including differences between smooth bathymetry and abyssal hills. Following this was a period of planning for the full-scale MODE-1 array, including an intensive "summer camp" at the NCAR in Boulder, Colorado. MODE-I was the major experiment spanning 5 months in 1973. Observations included (a) 26 subsurface moorings supporting 83 current meters and

50 temperature-pressure recorders, (b) 20 SOFAR floats, and (c) 12 repeated STD surveys over a 77-station grid with spacing from 33 to 50 km, all with approximately uniform coverage over an area roughly 200 km in diameter and augmented by several limited or special-purpose observations. The last MODE effort, POLYMODE, included long-term time series extending from MODE-1 and substantial geographical exploration using mooring arrays, improved SOFAR floats, and repeated closely spaced XBT sections.

One of the revolutionary aspects of MODE was the way it was designed. The summer camp to plan MODE-1 began with vigorous debates about the relative emphasis to be placed on improving the statistical description of the mesoscale (e.g., wavenumber–frequency spectra), developing a multivariable synoptic picture that could be compared with model simulations, and on testing dynamical principles (e.g., geostrophy, conservation of potential vorticity). Klaus Hasselman argued for a statistical description, which would drive observation placement toward a uniform co-array (collection of observation separations). Modelers favored an array that provided good coverage for initial and boundary conditions with some interior observations to verify model predictions. Bretherton and Chris Fandry arrived with an appreciation for how Gandin's (1965) objective analysis could help design sampling arrays while, coincidentally, I had unwittingly reinvented the same method in Fourier space. We (Bretherton, Davis, and Fandry, 1976) used objective analysis to address array measurement accuracy and came to appreciate the way that data-based maps depend as much on the statistics assumed in mapping as on the data and that the dynamics these maps imply are set mainly by the statistics. Consequently, dynamical tests are sometimes best done directly with statistics (see Hogg, 1974).

Eventually common sense held sway over sophisticated arguments. It was decided to accurately map as large an area as possible so that model verification, statistical description, and data-based dynamical testing could all be carried out. Extensive experimentation with many possible arrays and various statistical representations based on MODE-0 recommended a roughly uniform array of 200-km diameter. At the same time, Bill Schmitz at WHOI had designed an array to meet the same general objective based on a verbal report of the correlation scale and without calculation. I have always appreciated that the two approaches led to virtually identical arrays; objective analysis, however, could estimate mapping errors.

The ambitious MODE effort depended on emerging observation capabilities, but early results showed the limits of those capabilities. MODE-0 used both surface and intermediate moorings and an important comparison (Gould and Sambuco, 1975) showed that current meters on surface moorings had significantly elevated kinetic energy, reflecting errors caused by wave-induced mooring motion. It was known that in oscillatory flow the Savonius rotor accelerated faster than it decelerated, magnifying speed. New laboratory tests showed that the vane was too slow to completely reverse in oscillating flow, further magnifying speed, and, most seriously, vertical motion in the line below surface buoys caused rotors to speed up. After a decade of use, MODE-0 showed that these errors were so great as to make measured currents little more than qualitative. These findings and a disappointing data return from MODE-0 had important consequences. Bill Schmitz took a stronger hand in the Buoy Group and applied technical leadership in design, operations, and personnel; the improvements throughout MODE and after were significant and obvious. The Buoy Group also stopped using surface moorings except when they were needed to observe meteorology or the near-surface ocean. In short order subsurface deployments of 2 years became common.

An improved SOFAR float introduced by Rossby and Webb was a big step toward the duration and tracking range needed for basin-scale observations. These giant floats, 5 m long and weighing 425 kg, broadcasted at 270 Hz, could be tracked from 2000 km or more, and were designed to operate for a year. High mechanical stresses produced by the loud sound source caused initial mechanical failures in the MODE floats. High-frequency pingers allowed them to be recovered and as Doug Webb (personal communication) reports,

> The early failure . . . was caused by unbonding of the urethane window on the transducers. Fortunately R.V. Researcher was equipped with a large lathe. The floats were recovered, the faces removed, machined, a polyurethane disk inserted in a machined groove and an improvised 60 ton manually operated press deformed the aluminum to seal the face in place, and the floats were relaunched.

After repair the floats functioned well throughout MODE-1 and two were tracked for 2 years. The data awakened great interest in analysis of Lagrangian data and gave us the first comprehensive views of horizontal stirring and particle dispersion in the ocean (Freeland, Rhines, and Rossby, 1975). Unfortunately, the policy that publication rights belonged to data producers, led to many early and valuable studies encouraged for planning being relegated to the gray literature—the oft-cited fundamental analysis of Lagrangian statistics, "Particle dispersion in the western North Atlantic" by Jim Price (1982) of WHOI comes to my mind. Many simply abandoned the program because of this policy. Today, all participants in research programs typically can collaborate in publishing data analysis; the Tropical Pacific TOGA Observing System and Argo both distribute data immediately.

From the observational point of view, many of MODE's scientific payoffs were realized by statistical descriptions of scales, vertical modes, and lateral propagation and by tests of geostrophy and horizontal nondivergence carried out with statistics. The results most in keeping with the array design were the objective-mapping analyses by McWilliams (1976a, b) in which float velocities were combined with density from the STD/CTD surveys to produce maps of geostrophic streamfunction at all levels. Quasi-geostrophic potential vorticity was computed from these maps and its conservation examined as a test of quasi-geostrophy. McWilliams argued that (a) because MODE-1 was too short for statistical reliability, any dynamical test needed to be applied on a synoptic basis and (b) that a test of potential vorticity conservation was preferable to a direct test of geostrophy because it was less affected by inevitable observational error. The first point was Hasselman's argument for a statistical array put the other way.

Figure 4.3. A 5-m-long MODE-era SOFAR float with its developers, Tom Rossby (left) and Doug Webb (right), at its base.

A FOCUS ON AIR–SEA INTERACTION

After MODE, observational oceanography expanded with various collaborative efforts inside and outside the International Decade of Ocean Exploration. At the same time new observational groups were getting in action and new instruments were being developed. One of the thrusts was a new focus on air–sea interaction. At NOAA's Pacific Marine Environmental Laboratory (PMEL), David Halpern built a new mooring group to investigate air–sea interaction processes in the tropical Pacific, including the equatorial zone where El Niño was then known but not understood. Halpern had some significant advantages including expertise learned while working with the WHOI Buoy Group, new commercial acoustic releases and current meters, and one of the most talented technologists in the country, Hugh Milburn. He also had a daunting challenge of perception at least. In 1971 Bruce Taft and the WHOI Buoy

Group (Taft *et al.*, 1974) placed five surface moorings between 1°S and 1°N in the equatorial Pacific to observe current variability. Four of the moorings were lost, leading to the general perception that it was extremely difficult to maintain moorings in this region. Under Halpern's leadership, Milburn's careful quantitative design studies and innovative choice of components, including low-drag fairings on the mooring line in high-current regions, led to a first successful deployment of 1 month in 1976 (Halpern *et al.*, 1976) and an eventual service life of 6 months. This development opened the equatorial Pacific to sustained moored observations and, through the experience and testing gained, eventually led to the ATLAS moored buoy (Milburn and McLain, 1986) that was the mainstay of the supremely successful TAO array.

At the Scripps Institution of Oceanography (SIO), interest turned to air–sea interaction and the anomalies of North Pacific surface temperature brought into focus by Jerome Namias. The NORPAX program began in the late 1960s under the direction of John Isaacs and Tim Barnett with a focus on these anomalies, which today we associate with the Pacific Decadal Oscillation and remote responses to El Niño. This work had begun with extensive observations using tautly moored Bumble Bee buoys (famous for recording data by photographing panels of gauges). NORPAX planned to use Monster Buoys, which at 12-m diameter and 50 tons were aptly named buoys powered by an onboard diesel generator and reporting data by high-frequency radio links (Petre and Devereux, 1968). In the early 1970s the high cost of Monster Buoys and a tenuous science plan led ONR and NSF to reconstitute the program under the leadership of Charles Cox. Cox, renowned for his pioneering work in microstructure and much more of a creative genius than an organization man, was an inspired choice. Initially the focus stayed on the North Pacific. New academic participants were sought and new observational tools were instituted, particularly widespread use of XBTs and development of surface drifters and the shipboard Acoustic Doppler Current Profiler (ADCP).

No good deed goes unpunished. Warren White and Buzz Bernstein vigorously complained about Cox's suggestion that they create the TRANSPAC XBT program. Nevertheless, they built a tremendously successful observing system (see White, 1987) that produced a substantial data set and supported numerous publications, particularly by White and Bernstein. The program was eventually taken over by NOAA and languished as the number of probes deployed decreased each year.

A better-received suggestion from Cox was development of the shipboard ADCP. He learned of commercial acoustic backscatter "speed logs" used on ships and suggested they could provide substantial valuable data on upper-ocean currents if they could be range gated. ONR supported Lloyd Regier and I to work with Amatek-Straza and later RD Instruments to convert speed logs into ADCPs. For technical reasons associated with range gating and the difference between bottom reflection and volume backscatter, this was more complicated than adding range bins to the speed log. A major hurdle was developing a flexible yet accurate method of measuring the Doppler shift and in this respect the pioneering work on acoustic Doppler profilers by Pinkel (1979) was extremely important to the design discussion. We never really achieved the desired transparent processing scheme, but the RDI frequency-tracking circuitry did work. After several years of testing on Scripps ships, during which the demands on installation and on the compasses available on ships were discovered, the RDI shipboard ADCP was in operation. The first report of its use is Joyce, Wunsch, and Pierce (1986) and Chereskin *et al.* (1989) explains the foibles of the RDI frequency tracker.

The NORPAX surface-drifter effort turned out tragically. Drifters were attractive in NORPAX because the scientific focus was on the largest scales of variability near the surface and because satellite tracking was becoming feasible, albeit at a cost 10 times what it is today. Leaders in the NORPAX effort were Bill Richardson (the Buoy Project pioneer who had moved to Nova University) and Gerald McNally (SIO) who were tackling the problems of drogue and sensor survivability in satellite-tracked buoys as well as how winds and waves affected accuracy of drifter-measured surface currents. My own contribution was developing inexpensive drifters tracked by, and reporting surface temperature measurements through, high-frequency radio transmissions received by military direction-finding stations around the Pacific. Richardson's fundamentally experimental approach was appropriate to the key questions of longevity and our groups planned a wintertime trial in the Gulf of Maine using Richardson's R/V *Gulf Stream*. Wayne Hill, a bright and vigorous young engineer, and I were to join the trial in January 1975. For reasons now forgotten, at the last minute I could not join the ship. The first report was that the *Gulf Stream* was missing. Despite a massive Coast Guard search augmented by an aircraft and two tireless pilots from Nova University, only one crewman's body and some debris were ever found. The R/V *Gulf Stream* was lost with all hands.

The investigation that followed the *Gulf Stream's* loss put a new light on my father's adage on eyewitnesses. Reports of the ship being sighted in different places were wildly inconsistent with each other. Then came the rumors, from Soviet submarines to suicide pacts, which grew without apparent basis. Evidently in a crisis contradictory information is inevitable, rationality is an early victim, and conspiracy theories flourish. Fortunately, not everything of the NORPAX drifter program was lost. Gerald McNally continued the quest for surface drifters to map large-scale surface circulation and he succeeded (see McNally, 1981).

A direct outgrowth of the interest in air–sea interaction and the problems with rotor-vane current meters was the development of the Vector Measuring Current Meter (VMCM; Weller and Davis, 1980). It was bad luck that early current meters measured speed and direction rather than components of velocity or even the component of velocity along the direction of a current-following vane. In either of these alternate methods high-frequency noise can be filtered without significantly biasing the low-frequency velocity components. But if mooring-induced high-frequency effects add to the observed speed, processing cannot reverse the error. As the name indicates, the VMCM measures current components. The VMCM propellers, with a cosine response to the angle between its axis and the current, respond to vector components in a quasi-linear fashion. Nevertheless, when the oscillatory component is large compared with the mean, the inherent nonlinearity of the propeller leads to errors up to 10% associated with the sensor measuring water that has already been changed in the propeller's wake. The lesson I take from this is that in the presence of noise, linearity is a prerequisite to accuracy. The VMCM was first used on a surface mooring in the 1977 Mixed Layer Experiment (MILE) where comparison with VACMs showed the VMCM to be less susceptible to mooring motion (Davis *et al.*, 1981). For several years it was the mainstay of our mooring work.

Between a growing interest in equatorial currents and the development of the VMCM, SIO began to develop a small mooring group. Halpern's PMEL group was of great assistance in this, sharing with us the basics, clever new tricks developed by Milburn, and general procedures. This made our group a third-generation in the Buoy Group line. At the same time Bob Weller finished his Ph.D. and went to WHOI to work with Mel Briscoe, eventually heading the surface-mooring part of the Buoy Group. One of Weller's early successes was elaborating the VMCM into a three-axis current meter with which he produced a remarkable observational description of Langmuir circulation (Weller and Price, 1988). It was with dismay that we observed the Buoy Group search for continuing leadership and eventually dissolve into three loosely related enterprises. In two ways the Buoy Group was a victim of its own success. First, as performance improved and the lifetimes of subsurface moorings made long current records feasible, there was less reason to devote resources to improving moored measurements. Second, the effort that was required to make the multiuser group successful was tremendous and those asked to shoulder the leadership role began to feel the effort was not worthwhile.

A PRODUCTIVE MEETING

Ever since MODE, oceanographers can hardly escape organizational meetings, a great many of which are inefficient uses of time. For me, two exceptional meetings were the MODE summer planning session and a 1982 meeting called at NCAR by Bill Large and Pearn (Peter) Niiler to improve ocean observations. At the 1982 meeting a group of oceanographers interested in observing technology surveyed the status of the field and, in an unusual collaboration, devised a plan by which each development would get the attention it deserved. Surface-drifter improvements, including survivability, reduced cost, wind and other meteorological sensors, current-following accuracy, and thermistor chains, were assigned to various teams coordinated by Niiler and Large. This effort led to the CASID air–sea interaction buoys used in experiments like STREX (Large, McWilliams, and Niiler, 1986) and Ocean Storms (Large and Crawford, 1995) and to drifters optimized to accurately measure near-surface velocity and temperature (Niiler et al., 1987) that played a central role in TOGA and WOCE.

Doug Webb (co-developer of the SOFAR float) and I agreed on a plan of action for two projects: self-contained ADCPs and autonomous subsurface floats. I was then working with RD Instruments to develop a self-contained ADCP suitable for use on moorings and, at the same time, was studying if basin-scale subsurface general circulation, then the province of hydrographers and scattered float studies, could be measured by large numbers of floats that infrequently rose to the surface to be located by the same satellite tracking systems used by surface drifters. Bretherton (1980) had already outlined the sampling problem in averaging energetic mesoscale variability to extract mean flow and it looked feasible if the observation cost per year of float observation could be significantly reduced. Because satellite tracking and data relay would make a vertically cycling float autonomous from acoustic tracking networks and significantly reduce the cost of low-density sampling arrays, I suggested to Webb that such a float would be tremendously useful. We agreed that I would finish up the ADCP work and obtain funding for the float project while he began work on it.

The main problems to overcome in a self-contained ADCP were finding a suitable data recorder and fashioning acoustic beam patterns that were concentrated enough that they could be made to scatter primarily off weak volume scatterers rather than stronger discrete targets like mooring components or the surface. RDI considered an electronically steered directional array but settled on four mechanically shaded transducers. Data storage was through a streaming magnetic tape recorder with good energy efficiency. Considerable work and extensive tank testing went into adjusting the transducer beam patterns to have low side-lobes. Even though developed for SIO under an ONR grant, the prototypes were delivered elsewhere; Schott *et al.* (1993) report their first use. Similar units, operating at 300 kHz and using simple narrow-band Doppler processing, were delivered in time for use in the Ocean Storms experiment (D'Asaro, *et al.*, 1995). By chance, Ocean Storms included a wind event that resonantly generated strong inertial currents in the mixed layer that then propagated downwards (and laterally), providing a great example of what the ADCP could describe. Comparisons with VMCMs showed that, when ADCPs pinged infrequently enough to last 9 months (90 pings per 15-minute average), mooring motion would affect their accuracy as noise rather than the bias seen in rotor-vane meters. ADCP–VMCM differences were approximately 10% of the 10–30 cm/s ambient currents and 70% of this difference was noise. The noise was substantially stronger from an ADCP mounted in the surface-buoy bridle than from one in the mooring line at 115 m. With the exception of the "fish" problem encountered on TAO moorings (Plimpton *et al.*, 1997), the self-contained ADCP has proven versatile and accurate and has nearly replaced discrete current meters.

The autonomous float project started slowly. The main problem was to repetitively change float buoyancy so it could rise from its depth of neutral buoyancy to the surface for satellite communication and then return to depth. Webb initially concentrated on novel techniques for varying buoyancy including individual capsules of chemicals that would generate gas when exposed to seawater. We also experimented with a flapper valve mechanism to rectify wave-induced vertical motion into buoyancy at the surface, but failed to make it function. After 3 years of frustration and with concern for the continuation of ONR funding in the absence of progress, I convinced Webb to shift to a more conventional electric motor driven buoyancy pump such as

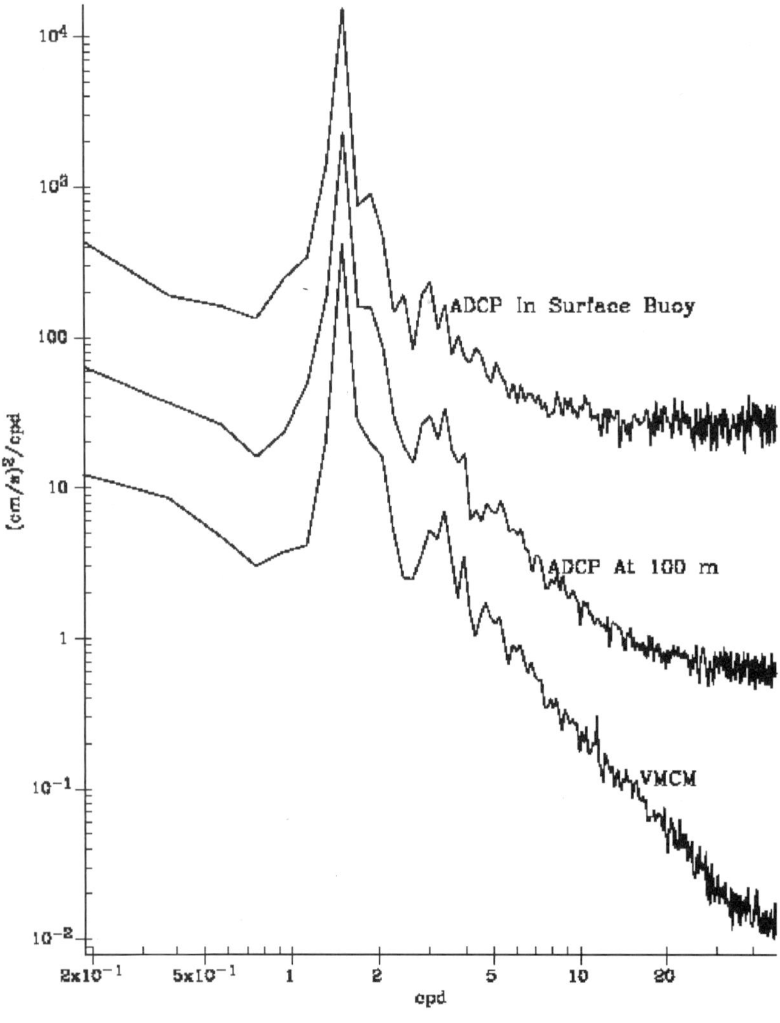

Figure 4.4. Frequency spectra of velocity at 60 m depth in the Ocean Storms as measured from surface moorings. (top) An ADCP mounted in the surface byoy; (middle) an ADCP mounted at 100 m in the mooring line; (bottom) a VMCM at 60 m. Motion of the surface buoy severely contaminates the observed signal. Similar, but weaker, contamination is seen from the 100 m ADCP. The VMCM, which records the integrated velocity over its 15-minute sampling period, is much less affected.

the one that Chip Cox (Duda *et al.*, 1988) had developed to repetitively profile microstructure. Webb had visited in France the Martec group where Claude Pacheco was working on an early version of the huge 300-kg multicycle buoyancy-driven float, Hippocampus. Its buoyancy engine was a small reciprocating Leduc hydraulic pump that pumped oil from inside the fixed-volume pressure case into a flexible external bladder, raising the vehicle's volume and buoyancy to ascend. Oil returned inside the pressure case to descend. Webb began developing a float using this buoyancy engine while, at the same time, SIO addressed the problem of keeping the antenna on a marginally buoyant vehicle above the surface in significant sea states.

WOCE

The World Ocean Circulation Experiment (WOCE) was the first oceanographic program to address the general circulation using global observations by the international community. The objective of WOCE was to significantly increase knowledge of the ocean's circulation by combining a global hydrographic survey, satellite measurements of sea-surface height, and global mapping of ocean velocity using drifters, floats, and high-density moored arrays all integrated by dynamical and inverse models. Float measurements near 1 km depth were to provide a level of known motion for referencing geostrophic shear maps, freeing inverse models to address key processes like turbulent mixing. The surface and float observations were the first sustained direct velocity observations taken to a global scale.

Three developments made possible the bold WOCE plan to directly measure circulation on two levels. First, Peter Niiler's (1987) group had developed a surface drifter that was shown to have minimal wind- and wave-driven slip through the water at its drogue. The design was unique from predecessors in two ways. The drogue is very much larger than the surface float, minimizing forces applied by winds and waves. This was made possible by using a small surface float that supported electronics and antenna for Argos tracking and a small intermediate float that in operation was usually submerged.

Second, the development by Tom Rossby and Don Dorson of the RAFOS float made feasible deployment of large arrays of acoustically tracked floats. The principle was simple: invert the SOFAR system (and name) using a few large sound sources and many relatively small and inexpensive quasi-Lagrangian floats that function as acoustic receivers. The floats record the times signals are received and, at the end of their life, pop to the surface to transmit these times through satellite. These 1.5-m-long floats have a mass of about 10 kg, simplifying deployment. The pressure case is a glass tube, minimizing manufacturing cost and matching thermal expansion to that of seawater. This, coupled with a spring-backed piston "compressee" to increase float compressibility to near that of seawater, gives the RAFOS float properties close to seawater so it approximately follows a water parcel. The initial sound sources were essentially moored SOFAR floats that could be deployed from volunteer ships but more efficient sources were soon developed. With a tracking range comparable to SOFAR floats, the low cost of RAFOS floats made it feasible to deploy large numbers of floats in an area and track them for times of order 1 year.

Third, the SIO–Webb collaboration led to development of the Autonomous Lagrangian Circulation Explorer (ALACE; Davis *et al.*, 1992). It was field tested

Figure 4.5. A modern RAFOS float with its developer Tom Rossby.

enough that in 1988 NSF supported further development and initial deployment of ten floats in the Southern Ocean at 750 m depth. Considering the many failure modes yet to be discovered, it was lucky that cycling every 15 days, four floats survived for 2 years and two traveled 110 degrees of longitude. This justified going ahead with WOCE plans to use ALACEs on the basin scales, but it also began a long battle for reliability, one that, much like the Buoy Group's early war against Murphy's Law, focused on myriad critical details. Because WOCE hydrographers generously agreed to deploy floats along the sections of the WOCE Hydrographic Survey, the cost of a measurement-year was basically the ALACE cost (about $10,000) divided by the years of service. With battery energy for 5 years, float reliability was the key to accurately describing the circulation. It was at this point that differences between the motives of Webb and Scripps began to grow. Webb was interested in innovation: the thermal powered float Claude; the bobber SOFAR float used successfully by Price (1996); and following up on Stommel's (1989) idea of the underwater glider Slocum. Scripps, particularly Jim Dufour and Jeff Sherman, invested effort in eliminating

Figure 4.6. A SOLO float in Argo configuration with the author (left) and Dean Roemmich (right), who created the Argo program.

design faults, developing manufacturing procedures that increased reliability, and buying parts from Webb to assemble floats in La Jolla. The average life gradually improved as problems were solved.

Float development continued during WOCE. When first learning of ALACEs, Terry Joyce (WHOI) suggested that they could be really useful if they measured profiles of temperature and salinity. Temperature profiles were added early in WOCE but suitable conductivity sensors were not available until 1995 and their initial performance was marginal. Stable conductivity measurements became routine when Sea Bird adapted their pumped conductivity sensor to floats for Webb Research, and subsequent uses have proven Joyce correct about interest in *T* and *S* profiles. Also in the mid-1990s, "vapor lock" of the hydraulic pump was identified as a persistent failure mode in ALACE. In this, the pump's cylinders become filled with gas that compresses rather than raising the internal pressure enough to open the exhaust valves. The source of the gas was apparently diffusion through the flexible bladders containing the oil that was pumped to change buoyancy. After failing to solve this problem, SIO began its own float design using a single-stroke hydraulic pump in which the cylinder was the container for the pumped oil when it was in the pressure case (this is the arrangement Cox used in his Cartesian Diver). Using some suggestions commissioned from Webb, the Sounding Oceanographic Lagrangian Observer (SOLO) was put in service for WOCE work in the North Atlantic. Davis *et al.* (2001) describe both Profiling ALACE and SOLO.

During WOCE, Michelle Ollitrault (IFREMER) and the group who first used a reciprocating-pump buoyancy engine in the Hippocampus float adapted that idea to build a multicycle RAFOS float called MARVOR. This design, first used in 1994, overcame the limitation of the original RAFOS concept that data was available only at the end of a float mission. MARVOR functions like a RAFOS float for a programmed time, rises to the surface to relay data, and then repeats the cycle until exhausted. The same design served as the basis for the autonomous profiling float PROVOR that is today used in Argo.

WOCE planning involved an exhausting series of meetings and reports in which the community tried to develop a coherent program. At a very early international meeting, I proposed using floats to map mean velocity on a single intermediate-depth level as a reference for hydrographic geostrophic shears. It was a tremendous concern that John Swallow questioned, in a gentle way, the utility of this: the widely separated floats would move incoherently, giving no idea of the structures leading to the statistical picture, and, cutting to the essence of Bretherton's (1980) analysis, variability could swamp the pattern of the general circulation. The float plan went ahead and when the South Pacific results began coming out, it was with great gratitude that I received Dr. Swallow's approval in the form of unpublished data from the Indian Ocean to use in planning that float array.

The various pre-WOCE developments contributed to the first direct-measurement descriptions of basin-scale circulation patterns. The Surface Velocity Program supported studies of variability, notably in the tropical Pacific; the mean global surface circulation; and, in conjunction with wind and altimetry measurements, the dynamics of the surface circulation (Niiler *et al.*, 2003). Acoustically tracked and autonomous profiling floats described intermediate-depth circulation in the subpolar North Atlantic and the North Atlantic Current (Lavender *et al.*, 2000; Bower *et al.*, 2002) and South Atlantic (Boebel *et al.*, 1999; Nuñez-Riboni *et al.*, 2005) while autonomous floats defined the intermediate-depth circulation of the Indian Ocean and South Pacific (Davis, 2005).

TODAY

Just as the WOCE float program was the direct inheritor of Swallow's pioneering work, so was the TOGA TAO Array the product of the effort begun by the WHOI Buoy Group. Development by Stan Hayes, Mike McPhaden, and the staff of the Pacific Marine Environmental Laboratory (PMEL) of the TAO Array (see McPhaden *et al.*, 1998) is one of the success stories of ocean observing. TAO reports equatorial atmospheric forcing, upper-ocean temperature, and some samples of salinity and currents in real time. Today TAO and its Japanese counterpart TRITON Array have been defined to be "operational" which, in the United States, means that day-to-day maintenance and improvement of TAO will be transferred from PMEL to an operational group. It will be an interesting experiment to see how well TAO evolves and maintains quality as it provides data for predicting ENSO after the scientists who motivated and built the array are removed from its operation.

In a complementary way, Dean Roemmich and John Gould led conversion of the WOCE autonomous float program into the internationally coordinated Argo program that today is two-thirds of the way toward its goal of maintaining an array of 3000 floats profiling salinity and temperature over the upper 2 km of the ice-free oceans. Because float operations are very different from mooring work, the structure of Argo is quite different from that of TAO/TRITON. Small groups in many nations (five in the United States) prepare and deploy floats with approved specifications while national centers and an international office see to the free distribution of the data. A loose organization assures that deployment plans are coordinated and that the array is maintained with approximately uniform coverage.

As the ocean observing system grows and becomes more operational, it will involve people who are not familiar with how ocean observations have evolved. It seems important to extract from history those lessons that will pertain to the more complex and operational observational systems we are moving toward. Above all, it seems certain that ocean observations will not soon become routine in the sense that people can be trained to sustain them without inventing solutions to new problems or improving measurement technology. Dedication, creativity, and skill were needed to achieve the progress of the last 50 years and these talents will be equally needed in the next 50 years.

While other areas like engineering, information sciences, and biology will produce new ideas and methods that can help ocean observations, the step of adapting them to the ocean will require special ocean expertise and creativity. Judging from history, there is a list of critical ingredients for successful technical development, including what will be needed to keep operational observing systems modern and efficient: time and monetary support; a degree of creative freedom and personal reward; good judgment; and a critical mass.

While Swallow created the first floats in less than a year, he continued to improve them for years. The Buoy Group took two decades to move from concept to operations. Autonomous floats were 8 years to the first serious deployment and 15 until the major bugs were worked out. Technologists will need time and a long cycle of testing and redesign to improve or even maintain the ocean observing system. Substantial funding will be needed to provide the required time.

No real progress is made in engineering or science without smart people being given the freedom to think in new and different ways, many of which will fail. The last 50 years were in this sense a golden age in oceanography. In research institutions and the small businesses that worked with them, individuals were motivated by various perceived personal rewards to try things and devise new methods. At first sight, it appears that the freedom and individual rewards that fuel creativity will be hard to sustain inside an increasingly operational observing system. This may be true, in which case we will end up more like the Post Office of 1980 than Federal Express of 2000, and the observing system will decline. But the experience at PMEL makes clear that delivering routine observations is not a routine undertaking and that the associated challenges can fuel the needed creativity.

Like all good engineering, successful ocean observing needs more than innovation. It also needs good judgment. It is well to always have a new, really imaginative idea but if it is observations we seek, good judgment is needed to temper these so that something comes of the best ideas. I am reminded that John Isaacs of SIO had myriad novel ideas from wave powered boats to skyhooks for lifting things into space, but in the end it was Richardson, Schmitz, and Milburn who set achievable goals, reached them, and methodically made moored observations practical.

Finally, it takes a community. As in the course of civilizations, progress occurs most rapidly when many people are involved in solving a problem and they communicate so that ideas can spread quickly. The history above is filled with collaborations that formed, dissolved, and were replaced by new combinations. In the same way, an idea in one area often gets moved to a new area where it unlocks an important path to progress. Even more obvious, some problems simply require a lot of different talents and consequently large diverse teams. We can only hope that a critical mass of talent and interest can be maintained in the next 50 years in the face of pressure to make ocean observation a routine task awarded to the low bidder.

ACKNOWLEDGMENTS

I am indebted to the many people who gave their time to help me get this wonderful story straight. Particular thanks go to Jerry Dean, John Gould, Jim Hannon, Nelson Hogg, Jim Luyten, Arnold Mantyla, Mike McPhaden, Jacky Moliera, Joe Reid, Doug Webb, and Bob Weller.

REFERENCES

Baker, D. J., 1981. Ocean instruments and experimental design. In: *Evolution of Physical Oceanography.* MIT Press, Cambridge, MA, 396–433.

Boebel, O., R. E. Davis, M. Ollitrault, R. G. Peterson, P. L. Richardson, C. Schmid, and W. Zenk, 1999. The intermediate depth circulation of the western South Atlantic. *Geophys. Res. Lett.* **26**, 3329–3332.

Bower, A. S., B. Le Cann, T. Rossby, W. Zenk, J. Gould, K. Speer, P. Richardson, M. D. Prater, and H.-M. Zhang, 2002. Directly measured mid-depth circulation in the northeastern North Atlantic Ocean. *Nature* **419**, 603–607.

Brainard, E. C., 1967. Evaluation of the P. O. E. Buoy with conventional buoy designs. In: *Transactions of the 2nd International Buoy Technology Symposium/Exposition.* Marine Technology Society, Washington, DC, 161–182.

Bretherton, F. P., 1980. Estimating the ocean general circulation using neutrally buoyant floats. Unpublished manuscript.

Bretherton, F. P., R. Davis, and C. Fandry, 1976. A technique for objective analysis and design of oceanographic experiments. *Deep-Sea Res.* **23**, 559–582.

Chereskin, T. K., E. Firing, and J. H. Gast, 1989. Identifying and screening filter skew and noise bias in acoustic Doppler current profiler measurements. *J. Atmos. Oceanic Technol.* **6**, 1040–1054.

Crease, J., 1962. Velocity measurements in the deep water of the North Atlantic. *J. Geophys. Res.* **67**, 3173–3176.

D'Asaro, E. A., C. C. Eriksen, M. D. Levine, P. Niiler, C. A. Paulson, and P. Van Meurs, 1995. Upper-ocean inertial currents forced by a strong storm. Part I: Data and comparisons with linear theory. *J. Phys. Oceanogr.* **25**, 2909–2936.

Davis, R. E., 2005. Intermediate-depth circulation of the Indian and South Pacific Oceans measured by autonomous floats. *J. Phys. Oceanogr.* **35**, 683–707.

Davis, R. E., and A. Acrivos, 1967a. The stability of oscillatory internal waves. *J. Fluid Mech.* **30**, 723.

Davis, R. E., and A. Acrivos, 1967b. Solitary internal waves in deep water. *J. Fluid Mech.* **29**, 593.

Davis, R. E., R. deSzoeke, D. Halpern, and P. Niiler, 1981. Variability in the upper ocean during MILE. Part I: The heat and momentum balances. *Deep-Sea Res.* **28A**, 1427–1451.

Davis, R. E., D. C. Webb, L. A. Regier, and J. Dufour, 1992. The Autonomous Lagrangian Circulation Explorer (ALACE). *J. Atmos. Oceanic Technol.* **9**, 264–285.

Davis, R. E., J. T. Sherman, and J. Dufour, 2001. Profiling ALACEs and other advances in autonomous subsurface floats. *J. Atmos. Oceanic Technol.* **18**, 982–993.

Duda, T. F., C. S. Cox, and T. K. Deaton, 1988: The Cartesian Diver: A self-profiling Lagrangian velocity recorder. *J. Atmos. and Oceanic Technol.* **5**, 16–33.

Freeland, H. J., P. B. Rhines, and H. T. Rossby, 1975. Statistical observations of the trajectories of neutrally buoyant floats in the North Atlantic. *J. Mar. Res.* **23**, 915–924.

Gandin, L. S., 1965. *Objective Analysis of Meteorological Fields*. Israel Program for Scientific Translation, Jerusalem, 242 pp.

Gould, W. J., 2005. From Swallow floats to Argo—The development of neutrally buoyant floats. *Deep-Sea Res. II* **52**, 529–543.

Gould, W. J., and E. Sambuco, 1975. The effect of mooring type on measured values of ocean currents. *Deep-Sea Res.* **22**, 55–62.

Halpern, D., J. R. Holbrook, H. B. Milburn, R. E. Davis, D. V. Hansen, and W. C. Patzert, 1976. Progress report of the EQUA-1 expeditions. *EOS, Trans. Am. Geophys. Soc.* **57**, 942.

Heinmiller, B., and N. Fofonoff, 1995. The WHOI Buoy Project: Mooring technology's early days. *Oceanus* **38**(1), 7–9.

Hogg, N., 1974: Balance of terms. In *Dynamics and analysis of MODE-1: Report of the MODE-1 dynamics group*. Department of Meteorology, Massachusetts Institute of Technology.

Joyce, T. M., C. Wunsch, and S. D. Pierce. 1986. Synoptic Gulf Stream velocity profiles through simultaneous inversion of hydrographic and acoustic Doppler data. *J. Geophys. Res.* **91**, 7573–7585.

Koblinsky, C. J., and N. R. Smith, 2001. *Observing the Oceans in the 21st Century*. Melbourne: GODAE Project Office. Bureau of Meteorology, 604 pp.

Large, W. G., J. C. McWilliams, and P. P. Niiler, 1986. Upper ocean thermal response to strong autumnal forcing of the northeast Pacific. *J. Phys. Oceanogr.* **16,** 1524–1550.

Large, W. G., and G. B. Crawford, 1995. Observations and simulations of upper-ocean response to wind events during the Ocean Storms Experiment. *J. Phys. Oceanogr.* **25**, 2831–2852.

Lavender, K. L., R. E. Davis, and W. B. Owens, 2000. Mid-depth recirculation observed in the interior Labrador Sea from subsurface floats. *Nature* **407**, 66–69.

McNally, G. J., 1981. Satellite-tracked drift buoy observations of the near-surface flow in the eastern mid-latitude North Pacific. *J. Geophys. Res.* **9**, 8022–8030.

McPhaden, M. J., A. J. Busalachi, R. Cheney, J. R. Donguy, K. S. Cage, D. Halpern, M. Ji, P. Julian, G.Meyers, G. T. Mitchum, P. P. Niiler, J. Picaut, R. W. Reynolds, N. Smith, and K. Takeuchi, 1998. The Tropical Ocean–Global Atmosphere (TOGA) observing system: A decade of progress. *J. Geophys. Res.* **103**, 14,169–14,240.

McWillimas, J. C., 1976a. Maps from the Mid-Ocean Dynamics Experiment: Part I. Geostrophic streamfunction. *J. Phys. Oceanogr.*, **6**, 801–827.
McWillimas, J. C., 1976b. Maps from the Mid-Ocean Dynamics Experiment: Part II. Potential vorticity and its conservation. *J. Phys. Oceanogr.*, **6**, 828–846.
MODE Group, The, 1978. The Mid-Ocean Dynamics Experiment. *Deep—Sea Res.* **25**, 859–910.
Niiler, P. P., R. E. Davis, and H. J. White, 1987. The water-following characteristics of a mixed layer drifter. *Deep-Sea Res.* **34**, 1867–1881.
Nuñez-Riboni, I., O. Boebel, M. Ollitrault, Y. You, P. Richardson, and R. Davis, 2005. General circulation of Antarctic Intermediate Water in the subtropical South Atlantic. *Deep-Sea Res. II* **52**, 545–564.
Petre, J. W., and R. Devereux, 1968. The Monster Buoy—A progress report for July 1967. In: *Marine Sciences Instrumentation, Vol. 4,* F. Alt, ed., Plenum Press, New York, 434–443
Pinkel, R., 1979. Observations of strongly nonlinear internal motions in the open sea using a range-gated Doppler sonar. *J. Phys. Oceanogr.* **9**, 675–686.
Plimpton, P. E., H. P. Freitag, and M. J. McPhaden, 1997. ADCP velocity errors from pelagic fish schooling around equatorial moorings. *J. Atmos. Oceanic Technol.* **14**, 1212–1223.
Price, J. F., 1982. Particle dispersion in the western North Atlantic. Unpublished manuscript.
Price, J. F., 1996. Bobber floats measure currents' vertical component in the Subduction Experiment. *Oceanus* **39**(1), 26.
Richardson, W. S., P. B. Stimson and C. H. Wilkins, 1963. Current measurements from moored buoys. *Deep-Sea Res.* **10**, 369–388.
Rossby, T., and D. Webb, 1970. Observing abyssal motion by tracking Swallow floats in the SOFAR channel. *Deep-Sea Res.* **17**, 359–365.
Rossby, T., and D. Webb, 1971. The four month drift of a Swallow float. *Deep-Sea Res.* **18**, 1035–1039.
Rossby, T., D. Dorson, and J. Fontaine, 1986. The RAFOS system. *J. Atmos. Oceanic Technol.* **4**, 672–679.
Schott, F., M. Visbeck, and J. Fischer, 1993. Observations of vertical currents and convection in the central Greenland Sea in the winter of 1988–1989. *J. Geophys. Res.* **98**, 14,401–14,421.
Stommel, H., 1955. Direct measurements of sub-surface currents. *Deep-Sea Res.* **2**, 284–285.
Stommel, H., 1957. A survey of ocean current theory. *Deep-Sea Res.* **4**, 149–184.
Stommel, H., 1989. The Slocum Mission. *Oceanography* **2**(1), 22–25.
Swallow, J. C., 1955. A neutrally-buoyant float for measuring deep currents. *Deep-Sea Res.* **3**, 74–81.
Swallow, J. C., 1971. The *Aries* current measurements in the western North Atlantic. *Philos. Trans. R. Soc. London Ser. A* **270,** 451–460.
Swallow, J. C., and L. V. Worthington, 1957. Measurements of deep currents in the western North Atlantic. *Nature* **179**, 1183–1184.
Taft, B. A., B. M. Hickey, C. Wunsch, and D. J. Baker, 1974. Equatorial Undercurrent and deeper currents in the central Pacific. *Deep-Sea Res.* **21**, 403–430.
The MODE Group, 1978. The Mid-Ocean Dynamics Experiment. *Deep-Sea Res.* **25,** 859–910.
Weller, R. A., and J. F. Price, 1988. Langmuir circulation within the oceanic mixed layer. *Deep-Sea Res.* **35**, 711–747.
Weller, R. A., and R. E. Davis, 1980. A vector measuring current meter. *Deep-Sea Res.* **27A,** 565–582.
White, W. B., 1987. The Volunteer Observing Ship XBT program in the Pacific Ocean. In: *Further Progress in Equatorial Oceanography*, E. J. Katz and J. M. Witte, eds., Nova University Press, Fort Lauderdale, FL, 87–95.

Chapter 5

On Reconciling Oceanic and Atmospheric Estimates of Surface Heat Fluxes—and on Validating Greenhouse Models

J. S. GODFREY

LEARNING TO BE AN OCEANOGRAPHER

I started my science career fairly conventionally by completing a Ph.D. in High Energy Physics. My subject matter related to tiny particles so exotic they do not occur naturally in the centre of the sun. By 1968 I was bored, and ready to jump. I applied successfully for an oceanography job advertised by an Australian research

J. S. Godfrey, retired in Tasmania.

laboratory called the Commonwealth Scientific and Industrial Research Organisation (CSIRO); I was given a year's Post-Doc at Harvard University, to learn what an ocean was. I had a wonderful time discovering how strange, counterintuitive, and challenging the physics of our oceans can be.

My new boss, Bruce Hamon, had discovered the eddies in the East Australian Current; he recognized that theoretical skills were needed to interpret what he had seen. He had somehow hired three people—George Cresswell, David Webb, and myself—to fill one job, a feat he never explained to us. He wanted to give us the feel of what he had found, so he taught us all how to create dynamic height maps using Nansen bottles. Latitudes and longitudes in those days came from the ship's officers' fixes on sun and stars. After a few days in cloudy conditions, in a slow ship in complex, rapid currents, these were often out by 60 miles, or more than an eddy diameter. A month's work at sea was often followed by several months ashore, unraveling what we had seen (and where). Recently I talked about this to some students, and about how such maps can now be made almost as well at a press of a button on the latest altimeter data. Their reaction told me that I was old enough to have been part of history for them, a rather sobering experience.

TROPICAL OCEANS–GLOBAL ATMOSPHERE

Stories filtered through to Cronulla (my new home, a suburb of Sydney) that the sea levels and wind systems of the entire tropical Pacific could change drastically from one year to the next, along with the exciting idea (Bjerknes, 1969) that this might involve a positive feedback between the ocean and atmosphere—the El Niño–Southern Oscillation (ENSO) phenomenon. Neville Nicholls, an Australian meteorologist, suggested that ENSO events might start in Indonesia, just to our north [despite the fact that the dramatic ENSO signal in Sea Surface Temperature (SST) occurred in the eastern Pacific, over 10,000 km away]. This suggested opportunity to an Australian oceanographer like me; and so it proved. The 10-year Tropical Ocean-Global Atmosphere (TOGA) program was being planned and then implemented. I was invited to join the international Scientific Steering Group of TOGA, which provided a wonderful chance to meet with meteorologists and oceanographers from around the world.

Peter Webster and Roger Lukas—the guiding lights of the TOGA scientific steering group, once it was under way—recognized that the basic 10-year monitoring strategy that had been adopted for TOGA needed to be supplemented by a more intensive field observation experiment. This was to be a 4-month study of what happens when Madden–Julian Oscillations (periods of several weeks of enhanced westerly wind stress, thought to be possible precursors of ENSO) moved eastward out of Indonesia. It was considered particularly important to quantitatively track the air–sea heat fluxes which caused SST to change. Webster and Lukas organized a successful appeal to the U.S. Congress to finance the massive U.S. component of this ambitious experiment. Eventually we all got in our ships and planes and onto our islands, and commenced measuring from November 1992 through February 1993.

LEARNING TO MEASURE, AND CHECK, HEAT FLUXES

Especially for oceanographers, the centrepiece of the Coupled Ocean–Atmosphere Response Experiment (COARE) was Bob Weller's IMET mooring, located at 2°S, east of Papua New Guinea. Three ships—R/V's *Wecoma*, *Moana Wave*, and our Australian ship *Franklin*—worked in its neighbourhood. On the bow of each ship meteorologists like Chris Fairall, Clayton Paulson, Steve Esbenson, and Frank Bradley undertook the deeply laborious task of accurately measuring all four components of heat flux, the two components of momentum flux, and precipitation. At the stern we oceanographers performed the slightly easier job of keeping the SeaSoars and ADCPs going, to map out the ocean temperature/salinity (T/S) and current structure in detail. All told, 22 instruments had to operate simultaneously at near-maximum precision without gaps to close the heat budget, even to the seemingly modest accuracy of 10 W/m^2 (about 5% of the average input of solar radiation, the biggest term in the net flux).

I will not forget the sight of five research vessels steaming together past the IMET buoy, to calibrate their meteorological measurements. Planes flew low overhead, to calibrate theirs. Two intercomparisons like this proved crucial to the eventual success of this part of COARE, because the differences in raw data among the platforms were pretty sobering (due to difficult problems like salt on humidity wicks, birds sitting on radiometers, and distortion of wind speeds by the ships' superstructure).

By the efforts of Frank Bradley and others involved in trying to obtain the most accurate air–sea fluxes possible, the COARE IMET mooring, the ships, and the aircraft produced a new level of accuracy. Intensive pre- and postcalibrations, dedicated in-the-field comparisons, and new methods paid off. Special Teflon screens around the humidity sensor on the IMET mooring let moisture in without allowing salt crystals to grow. Intensive study of radiometer performance resolved problems with factory calibrations; with sensitivity to solar heating; and with the possibility that sea birds would land on one of the radiometers (to be detected by examining the minute-by-minute data on the assumption that the bird can only sit on one radiometer at a time). The results from the IMET buoy itself were described by Weller and Anderson (1996).

It took several years to sort these basic problems out; but when this was done, the resulting heat fluxes from the bow of the ship matched the ocean heat uptake from measurements at the stern of the ship to within the 10 W/m^2 we had set ourselves (Feng *et al.*, 2000). The oceanographic part of the problem was nontrivial, since it involved using the divergence of ADCP currents around small boxes to assess upwelling rates, requiring very accurate knowledge of any misalignment of the ADCP. These issues were reviewed by Godfrey *et al.* (1998). We had problems on *Franklin*, but we also achieved this 10 W/m^2 accuracy using the same techniques and data from a later Indian Ocean cruise (Godfrey *et al.*, 1999).

A SERIOUS MISMATCH

Such results represented a major step forward towards resolving a very frustrating situation in oceanography: namely, the significant divide between (on the one hand) our ability to diagnose what the present state of the ocean is, and (on the other) our relative inability to quantify the forcing functions that cause this state to change. This is especially true of the net heat flux; and this quantity is crucial to simulating the SST, the key ocean parameter in coupled climate models.

We have many accurate tools for measuring the present state of the ocean. We can routinely measure T and S to accuracies like 3 parts in 100,000. Combining this with altimetry and Argo float data can and does provide a remarkably accurate, global picture of the state of the upper ocean, essentially in real time. The contrast between this "conventional" oceanography and the situation with heat fluxes is quite embarrassing. Typical net heat fluxes into the ocean might be 100 W/m^2 at seasonal peaks, with a long-term mean of perhaps 30 W/m^2; so a 10 W/m^2 accuracy is indeed modest when expressed as a fraction of the signal one is quantifying—and even 10 W/m^2 accuracy is a research quality result. It has long been known that different global net heat flux products—obtained using raw data collected by merchant ships over many years—differ from one another by several tens of watts per square meter over huge areas. Some carefully prepared products show an annual mean net heat flux of as much as 40 W/m^2, when averaged over the entire globe—enough to heat the top 100 m of the ocean by about 3°C/year! Evidently, these represent serious biases.

I will argue later in this essay that if we can at least partly resolve this terrible mismatch between our knowledge of the present ocean state, and of the forcings that cause it to change, it will help us greatly in diagnosing what greenhouse gas increases are doing to our climate. I will also argue that climate scientists will increasingly be called upon to do this, in a timely manner, at higher and higher spatial resolutions; and the financial issues involved will be very large. Before addressing this, however, I want to explain why I think we will be able to obtain heat fluxes of much better quality than those available till recently.

HOW CAN THE TOGA–COARE ACCURACY BE TRANSFERRED TO GLOBAL HEAT FLUX CLIMATOLOGIES?

On the face of it, it might seem impossible to achieve global heat fluxes to accuracies like 10 W/m^2. If we have so far verified this accuracy only at small spots in the ocean, after a very rigorous and expensive field campaign, how can we hope to get similar improvements routinely in places like the Antarctic, where ships hardly ever go? Yet recent analyses which make use of satellite data and numerical weather prediction model output suggest that we *may* already have a flux estimation technique which in combination with ship observations will allow us to approach this accuracy.

In the 1990s, a group of researchers at Southampton Oceanography Centre (SOC), England developed a new global air–sea flux climatology. The new SOC climatology utilised results from a careful examination of the raw data obtained by merchant ships in the North Atlantic. The data contained standard meteorological reports of air temperature, humidity, wind speed and direction, SST, and cloud cover, together with additional information regarding the sensors used to make these measurements. The sensor information allowed corrections to be developed for various observational biases, for example a warm bias in measured air temperatures due to solar heating of the ship superstructure. Subsequently, these corrections were applied where possible at an individual level (for the first time) to each of the 30 million meteorological reports contained in the Comprehensive Ocean–Atmosphere Dataset (COADS) which were used to obtain estimates of the various fluxes.

In short, the SOC researchers were doing for the merchant ship data the same kind of hard work that the COARE researchers did on their raw data (see Josey *et al.*, 1998, for a full description of the SOC climatology). When the SOC climatology was compared with fluxes from IMET buoys in the subduction region of the eastern North Atlantic, agreement to within 10 W/m^2 was achieved at 3 of 5 sites and to within 20 W/m^2 at the other sites (Josey *et al.*, 1999). However, a significant bias of order 50 W/m^2 was still evident in a comparison with a short-period buoy deployment in the western North Atlantic. Little high-quality comparison data were available elsewhere, but in the Arabian Sea, the SOC annual mean net heat flux differed from that obtained from an IMET mooring by 18 W/m^2 (e.g., Weller *et al.*, 1998); and with the IMET mooring in the COARE region by 10 W/m^2. All this looked very encouraging. However, although the ship corrections had a significant impact at a regional level, the global net heat flux with the SOC climatology still averaged to 30 W/m^2—i.e., the global bias evident in earlier products had not been solved; perhaps because there are vast areas in the Southern Hemisphere with virtually no merchant ship traffic, and hence no surface data. Ongoing research continues to target the various causes of this bias. An alternative approach, which uses the mathematical technique of inverse analysis with hydrographic estimates of the ocean heat transport as constraints, has recently led to a globally balanced version of the SOC climatology (Grist and Josey, 2003).

Meanwhile, satellite observations improved rapidly through the 1990s and early 2000s. These have the huge advantage, compared to merchant ship data, of coming from single sensors, many of which scan the globe at frequent intervals: the "data gap" problem is thus immediately much reduced. There are certainly difficult problems with radiometer drifts, and with the exact interpretation of what one is sensing; but salt cannot encrust satellite sensors, and no birds can sit on them. They offered the promise—given ingenuity, attention to detail, and luck—of providing a global raw data set of more homogeneous data quality than that from merchant ships.

One example comes with the use of scatterometers for measuring wind stresses. The improved quality of satellite-based wind stress estimates has been evident for some years to ocean modellers, through the improvements that had been achieved in simulating observed ocean circulation, simply by driving ocean models with the new wind stresses and comparing with available observations (e.g., Kessler *et al.*, 2003; Sengupta *et al.*, 2004; Xie *et al.*, 2001).

For the heat flux per se, the most useful satellite observations have so far come from SSMI (from July 1987 onward; Wentz, 1997). QuikSCAT has good resolution but is available only in recent years (from June 1999 onward). The wind speed of SSMI is very similar to that of QuikSCAT. Yu *et al.* (2004a) describe SSMI, and discuss these issues. The older, generally excellent estimates of SST from infrared radiation (e.g., Reynolds *et al.*, 2002) were also supplemented by microwave measurements of SST that can penetrate through cloud—crucial especially in the tropics, where cloudiness coincides with strong winds that often dramatically reduce SSTs. The ISCPP satellite product estimates the shortwave and infrared fluxes at the surface, by a combination of radiation measurements from a variety of satellites—some geostationary, some roving—and careful radiative calculations through the atmospheric column. Estimation of near-surface air humidity from satellites is more problematic, but satellite products are available which can be used with care. The 10-m air humidity derived from SSMI can be obtained from Chou *et al.* (2003).

Yu and her collaborators at Woods Hole have sought resolutions of the problems with the satellite products; they have then put these resources together, to create a new global ocean heat flux product. The satellite data proved not to be enough, due to regional biases associated with different radiometer drifts on different spacecraft, and especially due to inadequacies in air humidity. These authors adopted an objective analysis that combines satellite observations with model outputs of surface meteorology (Yu *et al.*, 2004a,b). When the annual and zonal mean of the resulting net heat flux is compared with the SOC product in the Atlantic, as a function of latitude, good agreement is obtained—especially in the Northern Hemisphere, where the sampling density of ship observations used to form the SOC climatology is highest (see Figure 5.1). Recently, comparison has also been made with all available IMET moorings, TAO and PIRATA buoys in the equatorial Pacific and Atlantic, respectively, and with ship reports from several research cruises. The results are encouraging (Yu, personal communication).

It of course remains to be seen how accurate this new product is, in the rest of the world oceans; and experience certainly suggests caution in assuming some new product is of universally good quality. Three tools are available to test this accuracy: one is the use of IMET moorings in several locations, chosen to represent climate conditions prevailing over large areas of ocean. A second is the use of Volunteer Observing Ships, equipped with IMET flux-measuring gear; these complement the good time resolution of the first tool, with good space resolution along ship tracks.[1]

[1] As an example of the need for high spatial resolution of surface fluxes, note that all merchant ship climatologies of wind stress completely missed an important, narrow meridional maximum of wind stress curl just north of the equator in the east Pacific, which shows up in satellite wind stress data. Inclusion of this maximum greatly improves the simulation of equatorial Pacific currents in ocean models (Kessler

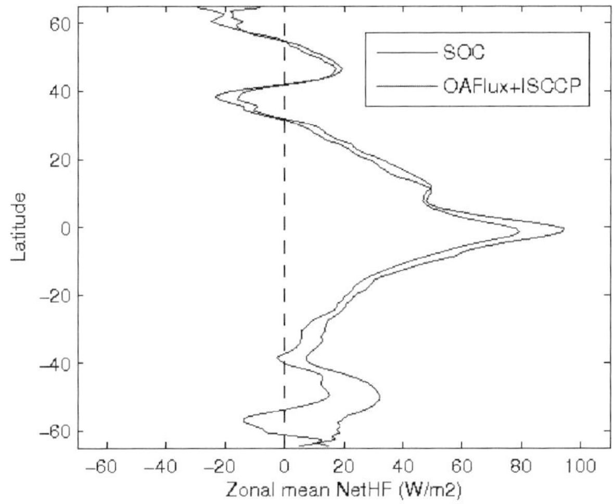

Figure 5.1. Comparison between the annual mean heat fluxes, averaged across the width of the Atlantic Ocean, as a function of latitude, from two climatologies: the Southampton (SOC) product, which is based on quality-controlled raw data from merchant ships, and the "OAFlux + ISCPP" product, which primarily depends on data from satellites, with some use of reanalyses of meteorological data. The SOC product has only marginally larger fluxes between 10 N and 50 N, and smaller fluxes everywhere else.

The third tool is the use of WOCE-style surveys of the ocean, to obtain heat divergence out of large areas of the ocean. Unfortunately, the original WOCE experiment was performed before the satellite data on which the WHOI flux product is based became available; but the WOCE methodology is well-known and can be used again.

A DIFFERENT ISSUE: HOW WILL WE FIRST FEEL GREENHOUSE WARMING EFFECTS ON OUR ECONOMY?

So much for the present status of measuring heat fluxes to a respectable accuracy, over the whole ocean. I now return to the question: are such (rather technical) developments important, from a societal point of view? I believe they are.

Over the last 15 years I have followed the evolution of ideas on greenhouse warming fairly closely, both as a scientist and as a citizen. My employer, CSIRO, is a conglomerate of many laboratories (divisions). Between them CSIRO covers a huge range of scientific expertise. In the late 1980s, the CSIRO Division of Atmospheric Research started leading an interdivisional study of greenhouse gas changes, and their likely effects on Australia. Nearly every year from then till 2002, I have attended the annual meetings of this group. They have been fascinating. Each year we hear of

---

et al., 2003). This fault in the older climatologies is presumably due to the fact that ship reports are made every 6 hours, during which time a ship can traverse this feature, typically 100 km wide. Continuous data from IMET-equipped ships should not be subject to such problems.

advances in tracking changes in trace gases—both in the atmosphere and from ice cores. We have watched three generations of global coupled models be built, each with higher resolution and better physics than the last; and we have seen marked improvements in simulating the present-day climate, in each new generation. We have heard results of various process studies—airplanes examining cloud nucleation, studies of gas emissions by cows, response of plants to increased CO_2, our own TOGA-COARE—and seen changes implemented in the models as a result. We have also seen this group of scientists respond to pretty vigorous criticisms, making changes based on those they judged to be valid, and simply wearing those they did not. I have not been a major contributor, which entitles me to say that I have been extremely impressed by the rigorous approach taken throughout this collective scientific effort.

This experience has provided one example of a phenomenon which I suspect we will see repeated many times in coming decades, as greenhouse warming rates increase. This phenomenon leads me to believe that development and verification of good ocean heat flux products—and work to make the ocean and atmospheric components of the coupled models also match observed ocean heat fluxes—should be a high-priority item for the climate science community. This phenomenon is as follows.

From 1911 to 1975, the net river inflow to the water supply of Perth, Western Australia fluctuated considerably on decadal time scales, but about a rather steady mean. In 1975 that inflow fell. As time went on, this looked less and less like an ordinary fluctuation and more like a long-term change. As of now, it looks like a step function; the average inflow from 1975 through 2004 is very close to half that from 1911 to 1975.

By the late 1990s, the Perth Water Board asked the Bureau of Meteorology and CSIRO to advise them: Is this due to natural climate variability? Or to greenhouse warming? Or to some other causes such as changing land use?

After a lengthy study of the output of nine global climate models with and without greenhouse gas increases, and "downscaling" statistical studies based on how observed rainfall at a point depended on large-scale weather patterns, the scientists involved gave their report. In essence they concluded that changes in land usage probably played a minor role. They said that climate variability played a major role, especially in the sharp onset of the drought. They noted that seven out of nine models showed Australian rainfall decrease (along with the warming!) to have a maximum in southwest Western Australia, as observed; though no model showed a drying rate nearly as fast as has been observed in that region. They did not offer an opinion on whether this meant that the drying was primarily due to natural variability, which for some reason mimicked the modelled greenhouse signal; or that there was some inadequacy (local or global), in all of the present generation of greenhouse models, to properly predict the large rate of drying in this region (which is small on a global scale). Almost inevitably, the scientists were caught in a difficult situation, since they were all well aware of the limitations—and huge complexity—of the models they were using to make an assessment, on which about half a billion investment dollars depended.

Assuming (as I do) that the advice given to the Perth Water Board was as good as could be expected given the present state of the coupled-modelling art, then I think this Western Australian experience is likely to be the first of many like it. I say this because all the greenhouse models agree on two things: (a) they all (from first generation through to the present ones) have predicted that climate warming should have already occurred since the 1970s, and most get the order of magnitude of the warming rate about right; and (b) the rate of warming is about to increase further, and to continue high for many decades into the future—even if we somehow drastically reduce greenhouse emissions. (The reason is that carbon dioxide has a lifetime of about 100 years in the atmosphere, so our present temperatures are a function of the last 100 years of CO_2 emissions.)

USE OF BETTER HEAT FLUXES TO VALIDATE GREENHOUSE COUPLED CLIMATE MODELS

If we regard the result (a) as a reason for taking the prediction (b) seriously, then I expect that in coming decades situations like that in Perth will crop up in many locations around the world, often simultaneously. We therefore need better tools for sorting out whether a decadal trend (in rainfall or evaporation, say) at a given location is part of a natural, interannual variation, or a harbinger of a more serious, long-term trend which may force major societal disruption at the place where it is occurring.

If the WHOI heat flux product—or improvements on it, based on a global verification program—proves to live up to its initial promise, I think it will prove an extremely valuable tool in such work. I will use the tropical Indian Ocean, as being the part of the ocean I have dealt with most in recent years, to illustrate what I mean.

One widely acknowledged weakness in global coupled models is their inability to predict the Asian monsoons well. At least as of 6 years ago, the atmospheric components of these models differed widely in where the monsoon rains fall; what its all-Asian average is; and what its interannual variability is, when forced by observed SSTs (Kang *et al.*, 2002). There are many reasons why this may be so. It may be due to issues such as representation of topography like India's Ghats mountains; or the fact that the atmospheric models do not get the (very large) diurnal cycle of rainfall at all accurately, suggesting some defects in underlying physics. These models also have great difficulty in simulating the observed strength and mean properties of the Madden–Julian Oscillations, referred to earlier.

One possible cause of such problems is that the mean heat fluxes they deliver to the ocean may be wrong (note that it is only the ocean that we need to be concerned about—the annual mean heat flux into land should be close to zero, since the heat cannot be mixed downwards or advected, as in the ocean). If the evaporative heat flux (the largest term in the tropics, after the incoming solar radiation) is wrong in a model, then the moisture transport towards rain areas—and therefore the rain quantity itself—must also be wrong.

By comparing observed and modelled heat flux climatologies, Godfrey *et al.* (submitted) and Yu *et al.* (submitted) have revealed two apparently very large problems: one with the ocean component, and the second with the atmospheric component, of the net surface heat fluxes in the two types of model that are coupled in most of today's climate forecasting work.

In two papers, we (Godfrey *et al.*, submitted) have compared observed heat flux climatologies with the results published so far, from ocean models, on the annual mean heat flux into the tropical Indian Ocean. The models' fluxes are all markedly too small—by a factor of about 2, if the WHOI climatology is correct. Two of these results come from a model we have run; our model has very coarse resolution, but we get similar net heat flux numbers to those from better models. In our own model we have tracked the problem down to numerical "dispersion," which causes the water column to go (spuriously) convectively unstable when upwelling rates exceed a critical value. The cause of this is well understood (e.g., Griffies *et al.*, 2000), and it is likely to occur in the other, higher resolution models. As a result, the absorption of heat into the ocean—whose spatial and interannual variations are the major cause of natural climate variability—are badly underestimated by the ocean component of coupled models, at least in the tropical Indian Ocean.

Yu and colleagues' results seem (on the face of it) to be still more drastic. According to the WHOI climatology, the Indian Ocean absorbs about 50 W/m^2 on annual and spatial average, north of 5°S. Two "reanalysis" products, which assimilate each day's meteorological data into a recent-generation atmospheric model, give heat fluxes of about 10 to 20 W/m^2 *out* of this region—i.e., these models seem to be asking the Indian Ocean to be a heat source for the atmosphere, rather than a sink as found in all the observation-based climatologies. I say "seem to be" because data assimilation usually involves the introduction of artificial heat and moisture sources into the model atmosphere, which will impact on its surface heat flux. A fairer test would be to examine the heat flux climatology of a model run over observed SST, with no data assimilation. Evidently, I am not qualified to comment on possible sources of the problem with the atmospheric models (if there indeed is a problem): but if Yu and colleagues' result is found also in such model runs, they at least will offer a big target for rectification efforts.

If the WHOI climatology lives up to its promise, I believe we first need to use it to identify errors in net heat fluxes between the individual ocean and atmosphere models and the observations, all over the world; and then (the hard part) to diagnose and hopefully correct the sources of error in the models. A well-designed effort of this kind should result in coupled models whose natural climatology is more realistic—and whose natural variability is also more realistic, probably on all time scales.

The new climatology would also provide authoritative data on long-term trends in heat fluxes into the ocean. For example, there is presently a massive trend of this kind in the tropical Indian Ocean (Yu *et al.*, personal communication), so big it is easily detectable in the present WHOI climatology. Testing coupled models for their skill in reproducing such long-term trends in heat fluxes will also provide a sophisticated and powerful tool for assessing the quality of greenhouse models.

Such a program is likely to occupy at least a decade, probably more. But coming back to the Perth water supply problem, and future problems like it: decisionmakers need to see proven success stories—"runs on the board"—from numerical models, before they will feel confident in making big financial decisions. The record of clear global greenhouse warming now extends back to the 1970s; it is extending in length—and probably also in intensity—decade by decade; we should, within a decade or two, be able to offer such decisionmakers more sophisticated analyses of what processes lie behind the observed warming, region by region. If we are able to explain to them, region by region, why certain climate trends have occurred in particular parts of the world over the last 30 years or so, they will be more willing to take action based on the model predictions for the next few decades.

However, if this is to come about, we *must* carefully check the promise of this new observational tool. This involves a long, expensive campaign of difficult observations, with IMET-type moorings and ship installations, WOCE-style hydrographic lines, and other such tools. It also presupposes that the satellite resources presently available are continued into the future.

Before closing, I would also like to add my voice to those protesting at the deeply wrongheaded policies being imposed on NASA. Their satellites have provided us with an unprecedented ability to keep track of the changes that are occurring to our climate—both natural and human-induced. Regardless of what one believes about greenhouse climate research, we need that ability into the future. We seem at risk of throwing such skills away, at a time when most climate scientists believe that greenhouse-induced climate change offers a "clear and present danger" to us all, in favour of putting one or two people on a piece of land much less attractive than any found on Earth. This is childish and irresponsible, and must not be allowed to happen.

ACKNOWLEDGEMENTS

I gave a draft of this essay to Bob Weller, Lisan Yu, and Simon Josey. They provided plenty of very constructive criticism of the technical aspects of the essay, which has allowed me to improve it greatly, in my own estimation. However, my brief was to write a "subjective essay," which is an invitation no retired scientist can refuse. I have indeed been subjective in places; I wish to state that all opinions expressed in this essay are strictly my own.

REFERENCES

Bjerknes, J., 1969. Atmospheric teleconnections from the equatorial Pacific. *Mon. Weather Rev.* **97**, 163–173.

Chou, S., E. Nelkin, J. Ardizzone, R. M. Atlas, and C. L. Shie, 2003. Surface turbulent heat and momentum fluxes over global oceans based on the Goddard satellite retrievals, Version 2 (GSSTF2). *J. Climate* **16**, 3256–3273.

Feng, M., R. Lukas, P. Hacker, R. A. Weller, and S. P. Anderson, 2000. Upper-ocean heat and salt balances in the Western Equatorial Pacific in response to the intraseasonal oscillation during TOGA COARE. *J. Climate* **13**, 2409–2427.

Godfrey, J. S., R. A. Houze, Jr., R. H. Johnson, R. Lukas, J.-L. Redelsperger, A. Sumi, and R. Weller, 1998. Coupled ocean–atmosphere response experiment (COARE): An interim report. *J. Geophys. Res. Oceans* **103**, 14,395–14,450.

Godfrey, J. S., E. F. Bradley, P. A. Coppin, L. F. Pender, T. J. McDougall, E. W. Schulz, I. Helmond, and T. J. Macdougall, 1999. Measurements of upper ocean heat and freshwater budgets near a drifting buoy in the Equatorial Indian Ocean. *J. Geophys. Res. Oceans* **104**, 13,269–13,302.

Godfrey, J. S., A. Schiller, R.-J. Hu, and R. Fiedler, 2005a. Exploration of the annual mean heat budget of the tropical Indian Ocean I: Studies with a coarse-resolution global model. *J. Climate* (submitted).

Godfrey, J. S., R.-J. Hu, A. Schiller, and R. Fiedler, 2005b. Exploration of the annual mean heat budget of the tropical Indian Ocean II: Studies with an idealized model. *J. Climate* (submitted).

Griffies, S. M., R. C. Pacanowski, and R. W. Hallberg, 2000. Spurious diapycnal mixing associated with advection in a z-coordinate ocean model. *Mon. Weather Rev.* **128**, 538–564.

Grist, J. P., and S. A. Josey, 2003. Inverse analysis of the SOC air–sea flux climatology using ocean heat transport constraints. *J. Climate* **16**, 3274–3295.

Josey, S. A., E. C. Kent, and P. K. Taylor, 1998. The Southampton Oceanography Centre (SOC) ocean–atmosphere heat, momentum and freshwater flux atlas. Southampton Oceanography Centre Report No. 6, Southampton, UK, 30 pp. & figs.

Josey, S. A., E. C. Kent, and P. K. Taylor, 1999. New insights into the ocean heat budget closure problem from analysis of the SOC air–sea flux climatology. *J. Climate* **12**, 2856–2880.

Kang, I., K. Jin, J. Kim, K. Lau, J. Shukla, V. Krishnamurthy, S. D. Schubert, D. E. Waliser, W. F. Stern, V. Satyan, A. Kitoh, G. A. Meehl, M. Kanamitsu, V. Y. Galin, A. Sumi, G. Wu, and Y. Liu, 2002. Intercomparison of atmospheric GCM simulated anomalies associated with the 1997/98 El Nino. *J. Climate* **15**, 2791–2805.

Kessler, W. S., G. C. Johnson, and D. W. Moore, 2003. Sverdrup and nonlinear dynamics of the Pacific Equatorial Currents. *J. Phys. Oceanogr.* **33**, 994–1008.

Reynolds, R. W., N. A. Rayner, T. M. Smith, D. C. Stokes, and W. Wang, 2002. An improved in situ and satellite SST analysis for climate. *J. Climate* **15**, 1609–1625.

Sengupta, D., R. Senan, V. S. N. Murty, and V. Fernando, 2004. A biweekly mode in the equatorial Indian Ocean. *J. Geophys. Res. Oceans* **109**, C10003.

Sun, B., L. Yu, and R. A. Weller, 2003. Comparisons of surface meteorology and turbulent heat fluxes over the Atlantic: NWP model analyses versus moored buoy observations. *J. Climate* **16**, 679–695.

Weller, R. A., and S. P. Anderson, 1996. Surface meteorology and air-sea fluxes in the western Equatorial Pacific warm pool during the TOGA coupled ocean–atmosphere response experiment. *J. Climate* **9**, 1959–1990.

Weller, R. A., M. F. Baumgartner, S. A. Josey, A. S. Fischer, and J. C. Kindle, 1998. Atmospheric forcing in the Arabian Sea during 1994–1995: Observations and comparisons with climatology and models. *Deep-Sea Res. Part II* **45**, 1961–1999.

Wentz, F. J., 1997. A well-calibrated ocean algorithm for special sensor microwave/imager. *J. Geophys. Res. Oceans* **102**, 8703–8718.

Xie, S.-P., W. T. Liu, Q. Liu, and M. Nonaka, 2001. Far-reaching effects of the Hawaiian Islands on the Pacific Ocean–atmosphere system. *Science* **292**, 2057–2060.

Yu, L., R. A. Weller, and B. Sun, 2004a. Improving latent and sensible heat flux estimates for the Atlantic Ocean (1988–1999) by a synthesis approach. *J. Climate* **17**, 373–393.

Yu, L., R. A. Weller, and B. Sun, 2004b. Mean and variability of the WHOI daily latent and sensible heat fluxes at in situ flux measurement sites in the Atlantic Ocean. *J. Climate* **17**, 2096–2118.

Yu, L., X. Jin, and R. A. Weller, 2005. Annual, seasonal, and interannual variability of the air–sea heat fluxes in the Indian Ocean. Submitted to *Journal of Climate* special issue on Climate Variability of the Indian Ocean.

Chapter 6

El Niño and Ocean Observations

A Personal History

MICHAEL J. McPHADEN

> When you can measure what you are speaking about and express it in numbers, then you know something about it.
>
> —Lord Kelvin, 1883

PROLOGUE

This article presents a brief personal account of my involvement in tropical ocean studies and observing system development over the past 30 years. The purpose is to provide perspective on the process of conducting scientific research against the backdrop of community wide efforts to better describe, understand, and predict El Niño. Failures as well as successes are highlighted to illustrate how unexpected events, blind alleys and supportive colleagues can shape a career. The most significant lesson from this experience is that systematic long-term observations are of fundamental importance to advancing our knowledge of the oceans and their role in regulating climate variability.

Michael J. McPhaden, Senior Research Scientist, Pacific Marine Environmental Laboratory.

INTRODUCTION TO OCEANOGRAPHY

As a graduate student at the Scripps Institution of Oceanography in the mid to late 1970s, the pursuit of tropical oceanography was for me a grand adventure. The field of tropical ocean dynamics, stimulated by Townsend Cromwell's discovery of the Equatorial Undercurrent in the Pacific in 1952 (Cromwell *et al.*, 1954), was still young and largely uncharted. The community of physical oceanographers interested in the tropics was small enough to know almost everybody by their first names.

People, events, and ideas influenced my research in ways I could never have anticipated on entering graduate school in 1974. I still recall the excitement of reading a preprint of Carl Wunsch and Adrian Gill's paper on the observational verification of 4- to 5-day-period equatorial inertia-gravity waves in the Pacific (Wunsch and Gill, 1976). Equatorial inertia-gravity waves were expected from theory (Moore and Philander, 1977) but until then had not yet been detected in the ocean. My initiation into research began with a theoretical investigation of these waves in the presence of mean flows like the Equatorial Undercurrent (McPhaden and Knox, 1979). I was fortunate that Bob Knox, an observationally oriented tropical oceanographer, and Myrl Hendershott, a theoretician, took an interest in supervising my research and cochairing my Ph.D. committee.

Other important early influences on my early development included sharing an office with fellow graduate student Jay McCreary who, in the process of completing his Ph.D. dissertation (McCreary, 1976), exposed me to new and exciting ideas about El Niño. The FGGE/INDEX/NORPAX[1] Equatorial (FINE) Workshop (Nova University, 1978), which brought together the leading equatorial theoreticians, modelers, and observationalists for 6 weeks of lectures and discussions in the summer of 1977 at Scripps, introduced me to Dennis Moore, Joel Picaut, Jim O'Brien, and many others whose names I had recognized from the literature but had not met before. Dennis Moore's Equatorial Theoretical Panel meetings in the late 1970s and early 1980s further nurtured my professional development by providing an informal but lively forum for exchanging ideas with the community of scientists interested in tropical ocean dynamics (see Chapter 7 in this volume). These meetings were rewarding not only because of the science that was discussed, but also because of the sense of collegiality they engendered and the lasting friendships they produced.

EL NIÑO

Understanding the ocean's role in El Niño and Southern Oscillation (ENSO) was an emerging research theme in the 1970s. The Norwegian-born meteorologist Jacob Bjerknes had recently identified the relationship between El Niño events and the

[1] FGGE, First Global GARP Experiment; GARP, Global Atmospheric Research Program; INDEX, Indian Ocean Experiment.

Southern Oscillation, a see-saw in atmospheric pressure between the eastern and western hemispheres first described by Sir Gilbert Walker in the early twentieth century (Bjerknes, 1966, 1969a). He also realized that El Niño involved the entire tropical Pacific basin, not just the coast of South America as was previously believed.

Bjerknes recognized that positive feedbacks between the ocean and atmosphere, mediated by variations in surface winds and sea surface temperature (SST), were critical to the generation of El Niño events. However, he did not understand what caused the trade winds to weaken at the onset of El Niño, or what processes shut down El Niño once underway. Likewise, his ideas about the physical oceanographic processes involved in El Niño were vague. He correctly ascribed unusually high SSTs in the eastern Pacific during El Niño primarily to a reduction in the intensity of equatorial upwelling; however, he believed incorrectly that the cause of the reduced upwelling was a weakening of the local trade winds.

Bjerknes was the first to link El Niño to patterns of weather variability over North America via atmospheric teleconnections emanating from the tropics. He also suggested that El Niño might be predictable and understood the need for systematic ocean observations in support of climate prediction. He expressed these ideas in particularly visionary remarks during a speech in 1969 (Bjerknes, 1969b):

> In the still farther future we can visualize the creation of a worldwide service of synoptic oceanography having as one of its most important duties to maintain monitoring buoys reporting by way of communication satellites such data which enter into the construction of transequatorial profiles at several selected geographical longitudes. That would usher in the era when attempts can be made to give electronic computers the right input for global long-range dynamical predictions of the fluctuations of the coupled circulations of the atmosphere and ocean.

El Niño research gained momentum with the 1972–73 El Niño when the Peruvian anchovy fishery, the largest fishery in world at the time, collapsed under the combined weight of overfishing and El Niño-induced environmental stress. Anchovies were widely used as feed supplement for poultry and livestock, so that collapse of the fishery (which accounted for nearly one-third of Peru's foreign exchange) rippled through the global economy (Glantz, 2001). In addressing the question of what physical processes controlled Peruvian coastal upwelling and therefore biological productivity during El Niño, Klaus Wyrtki identified the role of downwelling Kelvin waves[2] generated by relaxation of the trade winds in the central Pacific, rather than local wind variations in the eastern Pacific (Wyrtki, 1975). Shortly thereafter, the first dynamical theories for El Niño based on this concept began to appear (Hurlburt *et al.*, 1976; McCreary, 1976). El Niño research in the United States was further accelerated

[2] Kelvin waves are named after Lord Kelvin (Sir William Thompson), who first described these waves mathematically in the nineteenth century (Gill, 1982).

by the 1976–77 El Niño, which was linked to the extreme winter weather across the eastern half of the United States (Canby, 1977).

NORPAX AND EPOCS

The North Pacific Experiment (NORPAX) was established in the late-1960s with support from the U.S. National Science Foundation (NSF) and Office of Naval Research (ONR) to look for North Pacific influences on North American climate, building on ideas developed by Jerome Namais (1969). However, finding little convincing evidence from observations or atmospheric modeling studies that the North Pacific Ocean forced the overlying atmosphere to significantly affect seasonal climate variability over the United States (Chervin *et al.*, 1976; Davis, 1978), NORPAX looked south toward the equator to test oceanic aspects of Bjerknes's hypothesis for El Niño. A centerpiece of the NORPAX experiment was the Hawaii-to-Tahiti Shuttle Experiment, a sustained effort to study oceanic variability over a 16-month period during 1979–80 in the central Pacific between 152°W and 158°W (Wyrtki *et al.*, 1981).

The late 1970s also saw the establishment of NOAA's Equatorial Pacific Ocean Climate Studies (EPOCS) program to study El Niño and its climatic impacts over North America. EPOCS concentrated its activities in the eastern Pacific between 95°W and 140°W, a region where El Niño SST, thermocline depth, and sea level variations were large. An important oceanographic contribution of EPOCS was the establishment of long-term current meter mooring sites along the equator in the eastern Pacific to study changes in ocean circulation related to El Niño (Halpern, 1996).

First attempts by the Woods Hole Oceanographic Institution to anchor surface current meter moorings on the equator met with only limited success presumably because drag exerted by the strong vertically sheared South Equatorial Current and Equatorial Undercurrent parted mooring lines or submerged surface buoys (Taft *et al.*, 1974). In an effort to design a surface mooring that could survive in this severe environment, Hugh Milburn of NOAA's Pacific Marine Environmental Laboratory (PMEL) performed computer model experiments of taut line mooring dynamics in equatorial flow regimes to evaluate details of line lengths, system weights, and drag coefficients. These experiments led to a mooring design that incorporated plastic clip-on airfoil-shaped fairings to reduce drag in the areas of high current. PMEL successfully deployed one of these newly designed moorings on the equator at 0°, 150°W for 35 days in August–September 1976 (Halpern *et al.*, 1976) and a second mooring for almost 100 days in April–July 1977 at 0°, 125°W (Halpern, 1977). The success of these deployments demonstrated the feasibility of maintaining current meter moorings for long-term measurements along the equator where currents routinely exceeded 100 cm s^{-1} in the upper 200 m. Subsequent deployments during EPOCS established a 6-month design lifetime for equatorial moorings, after which mechanical wear and biofouling of the current meters became limiting factors.

EPOCS and NORPAX joined forces to coordinate the deployment of moored buoy arrays spanning 152°W (NORPAX) and 110°W (EPOCS) in the equatorial Pacific in 1979–80. Bob Knox, who was lead investigator at Scripps for the NORPAX moorings, temporarily enlisted my help to prepare mooring hardware as the time approached for field work to begin. This brief "hands on" experience was later rewarded with the satisfaction of having made a minor contribution to a major success story. The NORPAX/EPOCS moorings, combined with tide gauge measurements from the Galapagos Islands (91°W), detected the eastward propagation of a pulse in zonal currents and sea level along the equator over 60° of longitude in April–May 1980 (Knox and Halpern, 1982). This pulse propagated at the phase speed expected for a first baroclinic mode equatorial Kelvin wave and constituted the first compelling evidence for the existence of a key building block in Wyrtki's theory for El Niño.[3]

DRIFTING THERMISTOR CHAINS, PART I

My Ph.D. dissertation in 1980 was on theories of equatorial ocean circulation, including the effects of mean flows like the Equatorial Undercurrent on equatorial Kelvin waves. However, late in my graduate studies and as a postdoc at the National Center for Atmospheric Research (NCAR), I became interested in observational analyses and observing system development. Francis Bretherton and Eric Kraus entrained me in a feasibility design study for the "CAGE" experiment, an international effort intended to measure the heat balance of the North Atlantic from the top of the atmosphere to the bottom of the ocean (Bretherton *et al.*, 1984). The design study identified significant challenges, especially for estimating atmospheric heat flux divergences, which would make it difficult to compute the balance within specified error tolerances (Bretherton *et al.*, 1982). CAGE (not an acronym but intended to describe the volume over which measurements would be made) was thus never carried out, though it did influence later design of the World Ocean Circulation Experiment (WOCE).

As an offshoot of my involvement in CAGE, Eric Kraus introduced me to Jose Gonella of the Museum of Natural History in Paris, principal investigator of a drifting thermistor chain project as part of the Programme Français Océan et Climat dans l'Atlantique Equatorial (FOCAL). FOCAL was a French complement to the Seasonal Response of the Equatorial Atlantic (SEQUAL) program, a U.S. initiative to study the seasonal cycle of the tropical Atlantic over a 2-year period from 1982 to 1984. Jose was interested in developing a deployment strategy for the FOCAL drifters, and asked whether we could use the same techniques we developed for CAGE for this purpose. It was our opinion that we could and I volunteered to work on the problem. I subsequently spent March–April 1982 in Paris and Brest collaborating with French

[3] An El Niño did not develop in 1980 because there was no prior large-scale accumulation of excess upper ocean heat content along the equator, which is required in addition to the excitation of equatorial Kelvin waves for El Niño onset (Wyrtki, 1985; Cane *et al.*, 1986).

colleagues on a design study that considered alternative drifter deployment strategies targeted at improving estimates of the seasonal cycle in heat storage for the tropical Atlantic (McPhaden *et al.*, 1984). Guided by this study, we later deployed an array of 19 drifters, each equipped with a 117-m-long thermistor chain, between June 1983 and May 1984 (Reverdin and McPhaden, 1986).

The design study that guided the FOCAL drifting buoy deployment strategy was initiated coincident with the eruption of the Mexican volcano El Chichon and the stealthy onset of what would eventually be the strongest El Niño in 100 years. The 1982–83 El Niño proved to be not only a watershed event in the history of climate research, but also a major turning point in my career.

THE 1982–83 EL NIÑO

The 1982–83 El Niño, which was neither predicted nor even detected until nearly at its peak, caught the scientific community completely by surprise. At the time, most *in situ* oceanographic data collected were available only many months or, in some cases, years after collection. Some data on oceanic and atmospheric conditions from islands and volunteer observing ships were available in real time (within a day) or near real time (within a month for climate purposes), but they were far too few and scattered to be of much value in providing a coherent picture of evolving conditions. NOAA satellites capable of high-precision measurements of SST from space had been launched for the first time in 1981. However, unbeknownst to NOAA, SST retrievals after March 1982 were contaminated by stratospheric aerosols from the eruption of El Chichon. The aerosols produced a cold bias in the satellite SSTs that was mistakenly interpreted as clouds. These biased retrievals were flagged as bad and replaced with climatology in gridded analyses of the data. Thus, throughout much of 1982, SST analyses based on satellite data indicated near normal conditions in tropical Pacific. Moreover, those few *in situ* data that showed extraordinarily warm SSTs several degrees Celsius above normal were rejected as erroneous because they did not agree with the satellite analyses.

To complicate matters, the community at the time had an overly simplistic view of how El Niño evolved. Wyrtki (1975) had noted the tendency for the trade winds to strengthen and the sea level to build up in the western Pacific the year before El Niño. Rasmusson and Carpenter (1982) had also just published a composite of El Niño based on events from 1950 to 1973, which showed that typically anomalous warming progressed in stages from the west coast of South America in boreal spring to the central Pacific later in the year. None of these "canonical" developments occurred in 1982, lulling much of the community into the false belief that conditions in the tropical Pacific were near normal. Indeed, Klaus Wyrtki commented at a meeting of oceanographers and meteorologists at Princeton University in October 1982 that, "To call this an El Niño would be a case of child abuse!" It was only after scientists on a research cruise to the eastern Pacific reported that same month that the thermocline was 50–100 m deeper than normal did the community realize how badly it had been fooled (Toole and Borges, 1984). In terms of Lord Kelvin's famous quote, the sparsity of accurate oceanic measurements in 1982 exposed not only our ignorance about El Niño's complexity but also the gross inadequacy of existing observing systems to measure and describe it.

PLANNING FOR TOGA

The Princeton meeting had been called to lay the groundwork for U.S. involvement in a 10-year international effort to study El Niño, which eventually became known as the Tropical Ocean Global Atmosphere (TOGA) program (National Research Council, 1983). As a practical matter, it was clear that in the wake of the 1982–83 El Niño, plans would have to emphasize development of both El Niño observing and, if possible, forecasting capabilities. Thus, TOGA had three main objectives (WCRP, 1985):

1. To gain a description of the tropical oceans and the global atmosphere as a time-dependent system, in order to determine the extent to which this system is predictable on time scales of months to years, and to understand the mechanisms and processes underlying that predictability;
2. To study the feasibility of modeling the coupled ocean–atmosphere system for the purpose of predicting its variability on time scales of months to years; and
3. To provide the scientific background for designing an observing and data transmission system for operational prediction if this capability is demonstrated by coupled ocean–atmosphere models.

To get an early start on addressing the third of these objectives, NOAA's newly formed TOGA Project Office directed by Mike Hall convened a meeting at the Atlantic Oceanographic and Meteorological Laboratory (AOML) in Miami in May 1983 to discuss strategies for building a basin scale El Niño observing system in the Pacific Ocean (U.S. TOGA Project Office, 1988). Surface winds, sea surface temperature, upper ocean thermal structure, sea level, surface heat fluxes, and current velocity were identified as the key oceanic variables of interest. For each of these parameters, different measurement strategies were debated and prioritized. The focus was on *in situ* measurements vis-à-vis satellite measurements because satellite oceanography was still in its infancy and, except for sea surface temperature, no satellite missions for winds or sea level were operating or planned for the next several years. Also, satellites could not provide direct measurements of upper ocean thermal structure or currents, for which *in situ* data were essential. Real-time data relay to shore via satellite, where feasible, was viewed as a high priority to allow for routine monitoring of evolving climatic conditions and to support model-based prediction efforts.

The workshop arrived at an initial *in situ* observing system strategy to measure upper ocean thermal structure (Figure 6.1) based on methods that were considered

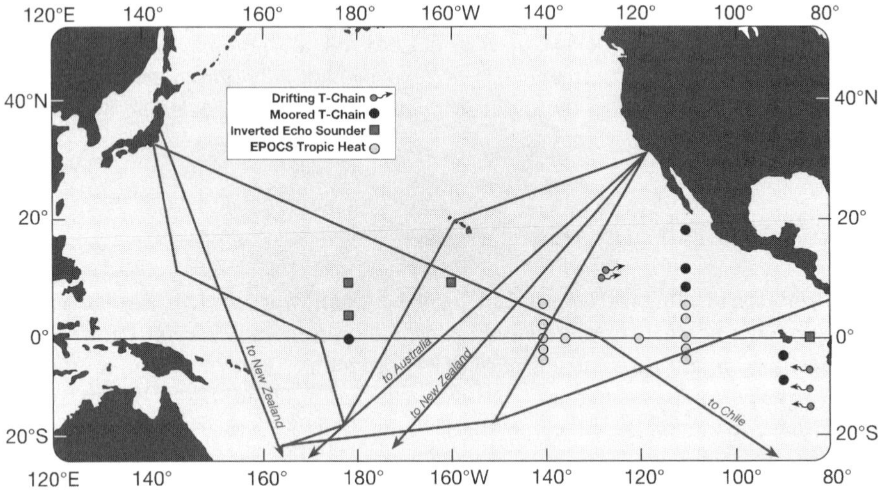

Figure 6.1. Initial design of a TOGA thermal field observing system. Solid lines indicate Ship-of-Opportunity XBT lines, EPOCS Tropic Heat indicates current meter moorings, and IES indicates inverted echo sounders, which provide a measure of sea level variability. Drifting and moored thermistor chains are also shown (after U.S. TOGA Project Office, 1988).

technically viable and economically feasible. Sampling characteristics of the proposed observing system components were viewed as complementary, but each had its limitations. For example, ship-of-opportunity expendable bathythermographs (XBTs), a proven and reliable technology, could provide good vertical (4 m) and along-track (1°) horizontal resolution. However, infrequent monthly temporal sampling contaminated the climate signals of interest with aliased energy from unresolved short-time-scale fluctuations, particularly near the equator. Also, XBT sampling was confined to fixed commercial shipping lanes, which left large areas of the ocean unsampled. Drifting thermistor chains, like those used in FOCAL or in the North Pacific for the 1980–81 Storm Transfer and Response Experiment (STREX) (Large *et al.*, 1986), could provide time series measurements at daily or higher temporal resolution in real time via Service Argos. However, they were subject to potential spatial sampling biases due to the effects of converging and diverging upper ocean currents. Moored thermistor chains, which were still under development, likewise potentially offered the advantage of high temporal resolution data in real time, but the presumed expense of moorings would limit their deployment to a few key regions. Thus, both drifting and moored thermistor chains were primarily viewed as "gap fillers" for the XBT network, which was considered the principal means of obtaining basin scale coverage for upper ocean thermal structure.

The AOML workshop recommended that a design study team be commissioned to systematically evaluate various sampling schemes for upper ocean thermal structure in the Pacific. In response, Bruce Taft and I convened a meeting in Seattle in July

1984 specifically to review relevant scientific issues, sampling requirements, existing methodologies, and next steps toward implementation of a basin scale thermal field observing system for TOGA (McPhaden and Taft, 1984). By this time, Stan Hayes had successfully advanced the development of a prototype moored thermistor chain and had scheduled deployment of a first mooring in the eastern equatorial Pacific for late 1984. Peter Niiler reported on a pilot deployment in the tropical Pacific of two drifting buoys with thermistor chains and five drifters with dummy chains of varying lengths to determine the effects of chain length on buoy motion. The workshop recommended that these efforts continue and expand as a complement to XBT sampling. In particular, the workshop concluded that, "Ultimately, we can anticipate deployments of about 10–12 moored thermistor chains . . ." and ". . . annual deployments of about 30 uniformly spaced drifters drogued with thermistor chains . . ." as part of the TOGA effort to routinely observe variations in thermal structure of the upper Pacific Ocean.

DRIFTING THERMISTOR CHAINS, PART II

In April 1985, Bill Large, Jim McWilliams, Peter Niiler, Bruce Taft, and I submitted a proposal jointly to NSF and the U.S. TOGA Project Office to support the development of a drifting thermistor chain project as part of the TOGA Observing System (the name ultimately given to the totality of systematic *in situ* and satellite measurement techniques used during TOGA). Large, McWilliams, and Niiler had previously carried out drifter thermistor chain measurements during STREX. Peter Niiler had already conducted initial engineering tests of drifting thermistor chains in the tropical Pacific. Bruce Taft was a longtime veteran of Pacific field programs and would coordinate logistics, deployments, and initial data processing from PMEL. I was on the research faculty at the University of Washington by this time (after sharing an office with Bill Large for awhile as a postdoc at NCAR) and had experience with the FOCAL drifting thermistor chain project. The proposal was for 3 years initially (1986–88) with the expectation that we would continue and expand if initial efforts were successful.

Our proposal reviewed well and we were awarded funds for an initial purchase of 10 drifters. We ordered from the same vendor that produced the STREX drifters, with a few modifications to the STREX design. One was that we required a 300-m-long thermistor chain rather than the 120-m-long thermistor chain used in STREX to ensure sufficient depth range to sample the upper thermocline throughout the tropical Pacific. Another was that we wanted the 12 subsurface thermistors on the chain multiplexed on three independent wire buses to mitigate data loss due to fish bite, which was a severe problem with the FOCAL drifters. In principle, neither of these modifications should have involved significant engineering challenges. In practice, though, they proved to be the Achilles heel of the project.

We had specified a design lifetime for the drifters of 1 year. However, our first installment of 10 drifters was poorly fabricated and delivered with software and hardware design flaws. A trial deployment of two drifters in the California Current in

July–August 1987 exposed a problem in which the temperature sensors failed progressively from the deepest to the shallowest depths. The manufacturer made design modifications to fix this problem, but sensors below 120 m systematically failed on a second trial deployment of four reconditioned drifters in the western tropical Pacific in February 1988. After another round of design modifications and upgrades to our remaining inventory, we staged a third field test in the California Current in July–September 1988 in which two drifting thermistor chains transmitted reasonably good data for 3 months. Encouraged by these results, we purchased 13 more drifters (3 with wind sensors) for a large-scale deployment in the western equatorial Pacific.

Nearly 2 years behind our original schedule, we launched 19 drifting buoys between 140° and 165°E in October–November 1989 with the intention of capturing the upper ocean response to a westerly wind burst. Our timing was perfect and within days of deployment we caught the onset of a strong wind burst (McPhaden *et al.*, 1992). However, the thermistor chains failed as they had in the first two field tests, with 50% of the sensors out of commission after only about 2 months (McPhaden *et al.*, 1991). Like the Neanderthal, drifting buoys with thermistor chains became an evolutionary dead end in the development of the TOGA Observing System.

ORIGINS OF THE TAO ARRAY

While the drifting thermistor chain project was frustrated by failure, Stan Hayes methodically advanced the development of the Autonomous Temperature Line Acquisition System (ATLAS) thermistor chain mooring at PMEL. Like Dave Halpern before him, Stan had a distinct advantage over the competition, namely, Hugh Milburn and his engineering division at PMEL. The ATLAS design incorporated many proven concepts from PMEL current meter moorings used in previous equatorial ocean studies (Milburn and McLain, 1986). In addition, significant cost savings were achieved by eliminating current meters in favor of temperature as the primary measurement. Also, elimination of current meters with their movable parts (rotors and vanes, or propellers) meant that mechanical wear and biofouling in the biologically productive upper equatorial ocean became less of a constraint on deployment duration. Thus, the ATLAS mooring design lifetime was 1 year, or twice that of a typical surface current meter mooring. Real-time data telemetry through Service Argos was also a standard ATLAS feature.

At the start of TOGA in January 1985 there was one ATLAS mooring in the equatorial Pacific at 2°S, 110°W transmitting surface air temperature and ocean temperatures in the upper 500 m. Stan added surface winds to the ATLAS measurement suite in 1986, adapting earlier design concepts developed for real-time wind measurements from current meter moorings (Halpern *et al.*, 1984). Initial ATLAS mooring deployments were concentrated between 110°W and 140°W as part of EPOCS (Hayes *et al.*, 1989). To extend the array into the western Pacific, Stan established collaborations with Joel Picaut in Noumea, New Caledonia, for deployments on French cruises along 165°E; and with Akimasa Sumi (University of Tokyo) and Kensuke Takeuchi (Hokkaido University) for deployments on Japanese cruises along 147°E. By the TOGA midlife review in Honolulu in July 1990, there were 17 ATLAS moorings deployed across the Pacific as part of the TOGA Observing System (Figure 6.2; McPhaden and Hayes, 1990; Hayes *et al.*, 1991). Stan named this array the TOGA Thermal Array in the Ocean (TAO) (Nova University, 1989).

The first half of TOGA witnessed significant progress on a number of other fronts as well. Between 1985 and 1990, the number of XBT lines regularly sampled in the tropical Pacific nearly doubled from 10 to 18, the number of TOGA Pacific tide gauges nearly doubled from 42 to 80, and surface drifters drogued at 15 m depth to measure mixed layer velocity and SST increased from only 8 to 164 (McPhaden *et al.*, 1998). Declassification of the U.S. Navy's GEOSAT satellite altimetry data made it possible to observe Kelvin wave variations associated with El Niño from space for the first time in 1986–88 (Miller *et al.*, 1988).

Early in TOGA, Cane *et al.* (1986) made the first successful prediction of an El Niño (the 1986–87 event) using a simple dynamical coupled ocean–atmosphere model. Two other research groups likewise issued successful El Niño forecasts during this time using statistical and statistical–dynamical hybrid models (Barnett *et al.*, 1988). Following the 1986–87 El Niño, unusually cold tropical Pacific SSTs in 1988–89 and their link to drought in the U.S. Midwest (Trenberth and Branstator, 1992) focused attention on the cold phase of ENSO, dubbed La Niña (Philander, 1990). New theories of the ENSO cycle between warm and cold events also emerged in which wind forced changes in thermocline depth, mediated by equatorial Kelvin and Rossby waves, were identified as the mechanism governing delayed negative feedbacks on the growth of tropical Pacific SST anomalies (Schopf and Suarez, 1988; Battisti and Hirst, 1989). The accumulation of excess upper ocean heat content at equatorial latitudes associated with these wave processes was also identified as a precondition for the occurrence of El Niño and as the source of predictability for ENSO time scale variations (Wyrtki, 1985; Cane *et al.*, 1986; Zebiak, 1989).

Building on studies prior to TOGA using simple "two-layer" ocean models (Busalacchi *et al.*, 1983), realistic wind-forced simulations of the El Niño were made for the first time from ocean general circulation models (e.g., Philander and Seigel, 1985). These modeling advances emphasized the need for accurate basin scale wind measurements to simulate ENSO-related changes in ocean circulation and SST (Harrison *et al.*, 1989). Also, NOAA's National Meteorological Center established the rudiments of an operational ocean data assimilation system for routine analyses of upper ocean thermal and current structures (Leetmaa and Ji, 1989). These analyses could be used to produce ocean initial conditions for coupled ocean–atmosphere model forecasts of El Niño and La Niña, but likewise depended on the availability of accurate surface wind forcing. Unfortunately, by 1990 it had become clear that the launch of the NASA scatterometer for high-precision surface wind measurements from space, anticipated for the second half of TOGA, would be delayed until after

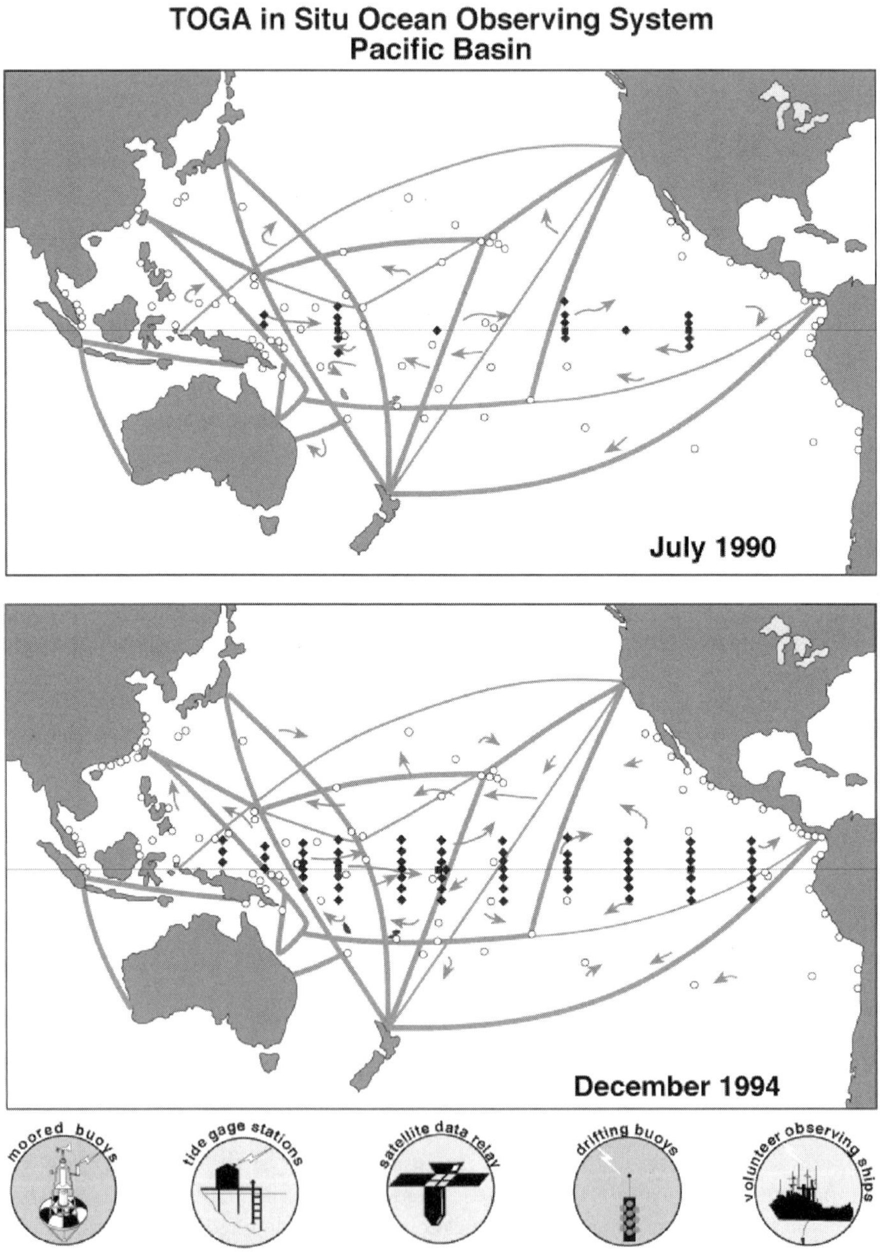

Figure 6.2. *In situ* components of the TOGA observing system at the time of the TOGA midlife review in July 1990 (top) and at the end of TOGA in December 1994 (bottom). The TAO array is indicated by solid black symbols representing ATLAS moorings (diamonds) and current meter moorings (squares) (after McPhaden *et al.*, 1998).

the program ended (National Research Council, 1990). Thus, in order for TOGA to meet its goals, enhanced *in situ* wind measurements would be needed in the Pacific.

It was against this backdrop of parallel developments in observing systems, theory, modeling, and forecasting that Stan Hayes boldly proposed a more than three-fold increase in the size of the ATLAS mooring array in the tropical Pacific during the second half of TOGA. Several factors made it possible to conceive of such a large-scale array: the technical success of early ATLAS deployments, the ability to measure both winds and ocean temperatures from the same mooring platform, the real-time telemetry of the data via satellite relay, and the much lower cost and longer design lifetime of the ATLAS mooring relative to traditional current meter moorings. Ocean general circulation model simulations and analysis of historical data provided specific design criteria for surface wind measurements that helped to justify the array (Harrison, 1989; Harrison and Luther, 1990). Stan astutely redefined the acronym from Thermal Array in the Ocean (TAO) to Tropical Atmosphere Ocean (TAO) array to capture the sense of expanded measurement priorities (Hayes *et al.*, 1991).

Both the U.S. and international TOGA scientific communities endorsed the concept of an expanded ATLAS array to address seasonal-to-interannual time scale climate studies (National Research Council, 1990; WCRP, 1991). At my first meeting as a new member of the International TOGA Scientific Steering Group in 1991, I can still recall the sense of excitement surrounding discussion of this plan. TAO was relatively big by the standards of previous *in situ* measurement programs undertaken in TOGA, but the community embraced this new idea with a surprising degree of unanimity and enthusiasm.

COMPLETION OF THE TAO ARRAY

In mid-1985, Stan had encouraged me to apply for a position recently vacated by Dave Halpern to lead the current meter mooring effort at PMEL. Despite the lack of prior experience in mooring programs, and the skepticism of one colleague who remarked that it was too much of a "right turn" in my career, I submitted an application, was ultimately selected, and joined PMEL in July 1986. While continuing to work on the then still apparently viable drifting thermistor chain project and other activities, I began a close and productive association with Stan that involved tight coordination of our ATLAS and current meter mooring programs. We also collaborated on nine journal articles over a period of 5 years making use of the moored data sets we collected. Stan was not only an excellent scientist and an effective leader, but also a pleasure to interact with on a personal level. Tragically, he passed away in 1992 at the age of 46 after a long and courageous battle with cancer.

With TAO unfinished and Stan gone, Eddie Bernard, Director of PMEL, asked me to assume responsibility for both the ATLAS and the current meter mooring programs. I was sufficiently familiar with the scientific objectives, operating procedures, and management structure that Stan had established, having worked with him for several years during the formative stages of TAO. The challenge initially was then to find enough time to do the work that two principal investigators previously had done. Highly competent technical staff and a supportive scientific community helped shorten the adjustment period, as did merging the ATLAS and current meter programs into a single unified program. Several of my colleagues and I also established a TAO Implementation Panel (TIP) under sponsorship of TOGA to help coordinate international contributions to the array.

The TAO array was completed in December 1994—the last month of TOGA—when the Taiwanese R/V *Ocean Researcher I* deployed an ATLAS mooring at 8°N, 156°E (McPhaden, 1995). The array took the full 10 years of TOGA to complete, with much of the growth occurring in the last 4 years of the program (Figure 6.2). Several institutions in the United States, Japan, France, Taiwan, and South Korea provided financial, technical, and logistic support. Full implementation required deployment of 401 moorings and 5.7 years of ship time (83 research cruises using 17 different ships from 6 different countries). When finished, TAO was called "the crowning achievement of TOGA" (Carlowicz, 1997) and far exceeded in scope what had been originally anticipated as a moored buoy component of TOGA (cf. Figures 6.1 and 6.2).

AFTER TOGA

TOGA achieved many of its research objectives thanks in large part to the synergy between observations, modeling, and theory (National Research Council, 1996). Data from the TOGA Observing System in particular fundamentally advanced our understanding of key oceanographic and meteorological processes involved in the ENSO cycle (McPhaden *et al.*, 1998), helped to foster advances in climate modeling (Stockdale *et al.*, 1998) and ocean data assimilation systems (Anderson *et al.*, 1996), and provided data necessary for initializing and verifying ENSO predictions (Latif *et al.*, 1998). In many respects, completion of the TOGA Observing System was a milestone analogous to the establishment of the World Weather Watch in the early 1960s to support numerical weather prediction (Davies, 1990).

It was recognized as early as the TOGA midlife review in 1990 that to critically evaluate the observing system for prediction purposes and to derive future benefit from its establishment, it would have to be continued after TOGA. However, TAO and other TOGA Observing System components had no guaranteed lifetime beyond the end of the program that sponsored them. As TOGA neared its end, therefore, the issue of sustaining the observing system became a matter of urgency. Several committees weighed in with recommendations for continuation, including the TOGA Scientific Steering Group (International TOGA Project Office, 1994) and the Ocean Observing System Development Panel (OOSDP), which was charged by the Intergovernmental Oceanographic Commission (IOC), the World Meteorological Organization (WMO), and the International Council of Scientific Unions (ICSU), to provide the conceptual design for long-term systematic ocean observations in support of climate (OOSDP,

1995). Likewise in the United States, Bob Knox chaired a committee of the National Research Council to consider the issue of long-term observations in support of seasonal-to-interannual climate prediction in the post-TOGA era. The "Knox Report" (National Research Council, 1994) concluded that there were strong arguments to sustain all components of the observing system that TOGA established, and that "the TAO array is of the highest priority for continuation."

It also became apparent that if TAO were to be sustained for the foreseeable future, a dedicated research vessel would be required to service the array. Ship time needs during the second half of TOGA were met using many ships from several nations, with logistics coordinated through the TAO Implementation Panel. These ship time arrangements were on a year-to year-basis, which worked reasonably well as a short-term strategy. However, a more stable base of ship support would be needed in the long term. Thus, NOAA acquired a surplus U.S. Navy submarine surveillance vessel in late 1993 in anticipation of using it for moored buoy maintenance work. The ship, the USNS *Titan*, was one of several that were no longer needed when the Navy mission changed with the fall of the Soviet empire and the end of the Cold War. After a conversion design phase, the ship was refit for buoy work in 1995 and rechristened in May 1996 as the NOAA ship *Ka'imimoana* (which means "Ocean Seeker" in Hawaiian). Commitment was further strengthened in November 1997, when the U.S. Congress authorized NOAA to provide $4.9 million per year in long-term support to cover the annual operating costs of TAO and other *in situ* components of the ENSO Observing System (which was the name given to the TOGA Observing System after TOGA ended).

TAO and other satellite and *in situ* elements of the ENSO Observing System were fully in place to capture the evolution of the 1997–98 El Niño, by some measures the strongest on record (McPhaden, 1999a,b; Picaut *et al.*, 2002). Unlike in 1982–83, the evolution of the 1997–98 El Niño was tracked day-by-day from its very beginning. TOGA advances in El Niño modeling and prediction, together with real-time data streams for model initialization, allowed forecasters to reliably anticipate some of the climatic consequences of the 1997–98 El Niño months in advance (Barnston *et al.*, 1999). It is estimated that in California alone, the economic benefits from advance warnings of El Niño impacts amounted to $1 billion (Changnon, 1999). In a postmortem assessment of the event, G.O.P. Obasi, Director General of the World Meteorological Organization, noted that "[t]he array of moored buoys stretching across the Pacific Ocean, originally established by the TOGA programme... has been an invaluable source of data for monitoring and modeling the event" (Obasi, 1998).

Aside from providing data in real time for ENSO monitoring and forecasting, TAO set a new standard for free and open access to oceanographic data for scientific research. From the start, TAO data were considered a community resource and made publicly available, generally within a day of collection. This policy represented a radical break from oceanographic tradition in which principal investigators would typically withhold new oceanographic observations from circulation for many months to years while they analyzed their data for publication. Easy access to and widespread use of TAO data, however, not only rapidly advanced communitywide scientific objectives, but also created vocal constituencies who advocated on behalf of the array's development, continuation, and later expansion.

TAO data have supported research efforts of scientists around the world, contributing to 30–50 publications per year in the refereed literature since the array was completed 1994. Most TAO-related publications have focused on seasonal-to-interannual time scale variability. However, the data have also supported studies of turbulent mixing, internal waves, and the diurnal cycle at one extreme of the frequency spectrum and decadal variability at the other extreme. As one example of scientific progress stimulated by the array, we have learned that equatorial Kelvin waves, whose first detection in the NORPAX/EPOCS moored time series data almost 25 years ago represented such a significant milestone, are a common feature of variability in the equatorial Pacific (Figure 6.3). These waves are most prominent at intraseasonal time scales (periods of 30–120 days) and are forced in the western Pacific by westerly wind bursts, the Madden–Julian Oscillation (Madden and Julian, 1994), and other forms of synoptic scale weather variability. Comprehensive basin scale observations of Kelvin waves have greatly improved our understanding of their dynamics and their influence on the development of El Niño events (Kessler, 2005).

Despite recent advances though, many questions remain unanswered about the nature of El Niño, La Niña, and the ENSO cycle between warm and cold events in the tropical Pacific. Specific dynamical linkages between intraseasonal atmospheric forcing and ENSO, the irregularity of the ENSO cycle, the limits of ENSO predictability, the decadal modulation of ENSO, and the impacts of global warming on ENSO are among some of the outstanding unresolved issues (McPhaden, 2004; van Oldenborgh *et al.*, 2005). Thus, far from being a solved problem, fundamental gaps in our knowledge about the ENSO cycle continue to challenge the scientific community today.

CONCLUSION

The TAO array provides one example of how research can guide the development of an ocean observing system for climate. It also demonstrates the feasibility of successfully sustaining an ocean observing system developed by research scientists over many years for climate purposes. In particular, TAO and its antecedent programs (e.g., EPOCS and NORPAX) span nearly three decades. The 0°, 110°W time series, originally started during EPOCS in 1979 and later continued as part of TAO, is now the longest moored time series in the world ocean.

The TAO array has continued to evolve over time, taking advantage of advances in scientific understanding, new measurement technologies, and the Internet revolution for data display and dissemination (http://www.pmel.noaa.gov/tao/). Partnerships to maintain the array have likewise evolved with time. TAO is now

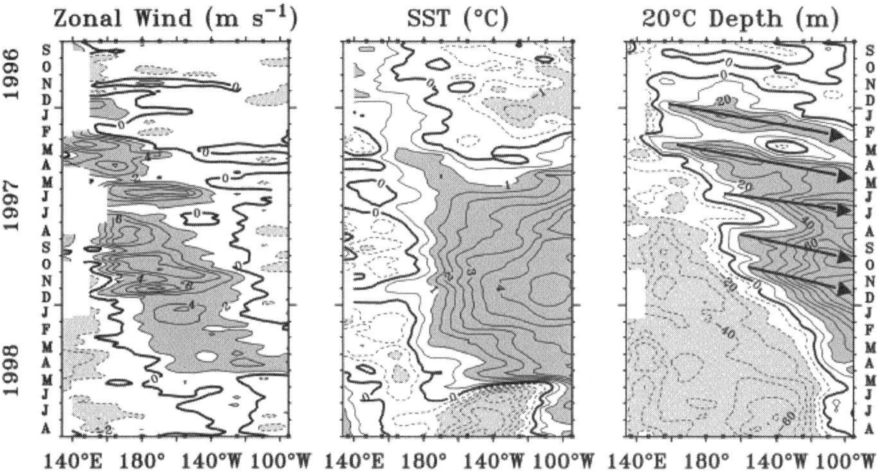

Figure 6.3. Time versus longitude sections of anomalies in surface zonal wind (left), SST (middle), and 20°C isotherm depth (right) from September 1996 to August 1998. Analysis is based on 5-day averages between 2°N and 2°S of moored time-series data from the TAO array. Positive winds are anomalously westerly, positive SSTs indicate unusually warm conditions, and positive 20°C isotherm depths indicate a deeper than normal thermocline. Arrows superimposed on the 20°C isotherm trace eastward propagating Kelvin waves excited by westerly wind bursts prior to and during the 1997–98 El Niño. These waves contributed to the development of the El Niño by progressively deepening the thermocline in the eastern Pacific. After McPhaden (1999a).

primarily supported as a bilateral effort between NOAA and the Japan Marine Earth Science and Technology Agency (JAMSTEC) and was renamed TAO/TRITON in January 2000 to recognize the contribution of JAMSTEC TRITON (TRIangle TransOcean buoy Network) moorings in the western Pacific.

The array has been designated an initial component of the Global Ocean Observing System (GOOS) by virtue of its proven scientific value and cost-effectiveness for climate studies (Nowlin *et al.*, 2001). A similar but smaller scale array has been implemented in the Atlantic Ocean to address ocean–atmosphere interactions associated with tropical Atlantic climate variability (Servain *et al.*, 1998). Plans also exist for an expansion of the moored array into the Indian Ocean to support research and forecasting related to the Asian-Australian monsoons and monsoon–ENSO interactions. These efforts represent significant steps toward realizing Bjerknes's vision of "... a worldwide service ... to maintain monitoring buoys reporting by way of communication satellites ... for long-range dynamical predictions. ..."

In conclusion, to echo Lord Kelvin, we can now measure El Niño and because of that we know something about it. This achievement represents a decades long community-wide effort. Sustaining and building upon this success are the next great challenges in global ocean observing system development for the 21st century.

ACKNOWLEDGMENTS

I thank Russ Davis, Paul Freitag, Markus Jochum, Bob Knox, Hugh Milburn, Dennis Moore, and Andy Shepherd for comments on earlier versions of this manuscript.

REFERENCES

Anderson, D. L. T., J. Sheinbaum, and K. Haines, 1996. Data assimilation in ocean models. *Rev. Prog. Phys.* **59**, 1209–1266.

Barnett, T. P., N. Graham, M. A. Cane, S. Zebiak, S.Dolan, J. J. O'Brien, and D. Legler, 1988. On the prediction of the El Niño of 1986–87. *Science* **241**, 192–196.

Barnston, A. G., A. Leetmaa, V. E. Kousky, R. E. Livezey, E. A. O'Lenic, H. Van den Dool, A. J. Wagner, and D. A. Unger, 1999. NCEP forecasts of the El Niño of 1997–98 and its U.S. impacts. *Bull. Am. Meteorol. Soc.* **80**, 1829–1852.

Battisti, D. S., and A. C. Hirst, 1989. Interannual variability in a tropical atmosphere–ocean model: Influence of the basic state, ocean geometry and nonlinearity. *J. Atmos. Sci.* **46**, 1687–1712.

Bjerknes, J., 1966. A possible response of the atmospheric Hadley circulation to equatorial anomalies of ocean temperature. *Tellus* **18**, 820–829.

Bjerknes, J., 1969a. Atmospheric teleconnections from the equatorial Pacific. *Mon. Weather Rev.* **97**, 163–172.

Bjerknes, J., 1969b. Global ocean–atmosphere interactions. Proc. Symposium on Physical Variability in the North Atlantic Ocean, International Council for the Exploration of the Sea, Dublin, Ireland, 25 September 1969, 538–549.

Bretherton, F. P., D. M. Burridge, J. Crease, F. W. Dobson, E. B. Kraus, and T. H. Vonder Haar, 1982. The CAGE Experiment: A feasibility study. World Climate Research Programme, Report WCP-22, Geneva, Switzerland, 95 pp.

Bretherton, F. P., M. J. McPhaden, and E. B. Kraus, 1984. Design studies for climatological measurements of heat storage. *J. Phys. Oceanogr. 14*, 318–337.

Busalacchi, A. J., K. Takeuchi, and J. J. O'Brien,1983. Interannual variability of the equatorial Pacific—Revisited. *J. Geophys. Res.* **88**, 7551–7562.

Canby, T., 1977. The year the weather went wild. *Natl. Geogr.* **152**, 798–830.

Cane, M. A., S. C. Dolan, and S. E. Zebiak, 1986. Experimental forecasts of the 1982/83 El Niño. *Nature* **321**, 827–832.

Carlowicz, M., 1997. Review of TOGA finds new culture, new puzzles. *EOS, Trans. Am. Geophys. Union* **78**, 4–5.

Changnon, S. A. 1999. Impacts of 1997–98 El Niño generated weather in the United States. *Bull. Am. Meteorol. Soc.* **80**, 1819–1827.

Chervin, R., W. Washington, and S. Schneider, 1976. Testing the statistical significance of the response of the NCAR general circulation model to North Pacific Ocean surface temperature anomalies. *J. Atmos. Sci.* **33**, 413–423.

Cromwell, T., R. B. Montgomery, and E. D. Stroup, 1954. Equatorial undercurrent in the Pacific revealed by new methods. *Science* **119**, 648–649.

Davies, A. (ed.), 1990. Forty years of progress and achievement: A historical review of the WMO. Rep. WMO-721, World Meteorological Organization, Geneva, Switzerland, 203 pp.

Davis, R. E., 1978. Predictability of sea level pressure anomalies over the North Pacific Ocean. *J. Phys. Oceanogr.* **8**, 233–246.

Gill, A. E., 1982. *Ocean–Atmosphere Dynamics*. Academic Press, Orlando, FL, 662 pp.

Glantz, M. H., 2001. *Currents of Change: Impacts of El Niño and La Niña on Climate and Society*. Cambridge University Press, Cambridge, 252 pp.

Halpern, D., 1996. Visiting TOGA's past. *Bull. Am. Meteorol. Soc.* **77**, 233–242.

Halpern, D., 1977. Current and temperature measurements in the Pacific Equatorial Undercurrent. *EOS, Trans. Am. Geophys. Soc.* **58**, 1155.

Halpern, D., J. R. Holbrook, H. B. Milburn, R. E. Davis, D. V. Hansen, and W. C. Patzert, 1976. Progress report of the EQUA-1 expedition. *EOS, Trans. Am. Geophys. Soc.* **57**, 942.

Halpern, D., H. P. Freitag, and A. Shepherd, 1984. Applications of Argos and RAMS measurements in equatorial Pacific ocean–atmosphere interaction studies. Proceedings of Oceans' 84, Marine Technology Society, Washington, DC, 736–740.

Harrison, D. E., 1989. Local and remote forcing of ENSO waveguide response. *J. Phys. Oceanogr.* **19**, 691–695.

Harrison, D. E., and D. S. Luther, 1990. Surface winds from tropical Pacific islands: Climatological statistics. *J. Climate* **3**, 251–271.

Harrison, D. E., W. S. Kessler, and B. Giese, 1989. Hindcasts of the 1982–83 El Niño: Thermal and dynamic variability along the ship of opportunity XBT tracks. *J. Phys. Oceanogr.* **19**, 397–418.

Hayes, S. P., *et al.*, 1989. The Equatorial Pacific Ocean Climate Studies (EPOCS) plans: 1986–1988. *EOS Trans. Am. Geophys. Union* **67**, 442–444.

Hayes, S. P., L. J. Mangum, J. Picaut, A. Sumi, and K. Takeuchi, 1991. TOGA TAO: A moored array for real-time measurements in the tropical Pacific Ocean. *Bull. Am. Meteorol. Soc.* **72**, 339–347.

Hurlburt, H. E., J. C. Kindle, and J. J. O'Brien, 1976. A numerical simulation of the onset of El Niño. *J. Phys. Oceanogr.* **6**, 621–631.

International TOGA Project Office, 1994. The TOGA Scientific Steering Group, Report of the 13th Session, Boulder, CO, 29 August–1 September 1994. International TOGA Project Office, Geneva, Switzerland, 13 pp.

Kessler, W. S., 2005. The oceans. In: *Intraseasonal Variability in the Atmosphere–Ocean Climate System*, K. M. Lau and D. E. Waliser, eds., Springer, Berlin, 175–222.

Knox, R., and D. Halpern, 1982. Long range Kelvin wave propagation of transport variations in Pacific Ocean equatorial currents. *J. Mar. Res.* **40** (Suppl.), 329–339.

Large, W. G., J. C. McWilliams, and P. P. Niiler, 1986. Upper ocean thermal response to strong autumnal forcing in the Northeast Pacific. *J. Phys. Oceanogr.* **16**, 1524–1550.

Latif, M., D. Anderson, T. Barnett, M. Cane, R. Kleeman, A. Leetmaa, J. O'Brien, A. Rosati, and E. Schneider, 1998. A review of the predictability and prediction of ENSO. *J. Geophys. Res.* **103**, 14,375–14,394.

Leetmaa, A., and M. Ji, 1989. Operational hindcasting of the tropical Pacific. *Dyn. Atmos. Oceans* **13**, 465–490.

Madden, R. A., and P. A., Julian, 1994. Observations of the 40–50 day tropical oscillations—A review. *Mon. Weather Rev.* **122**, 814–837.

McCreary, J., 1976. Eastern tropical ocean response to changing wind systems, with application to El Niño. *J. Phys. Oceanogr.* **6**, 632–645.

McPhaden, M. J., 1995. The Tropical Atmosphere–Ocean array is completed. *Bull. Am. Meteorol. Soc.* **76**, 739–741.

McPhaden, M. J., 1999a. Genesis and evolution of the 1997–98 El Niño. *Science* **283**, 950–954.

McPhaden, M. J., 1999b. El Niño: The child prodigy of 1997–98. *Nature* **398**, 559–562.

McPhaden, M. J., 2004. Evolution of the 2002–03 El Niño. *Bull. Am. Meteorol. Soc.* **85**, 677–695.

McPhaden, M. J., and S. P. Hayes, 1990. Moored velocity, temperature and wind measurements in the equatorial Pacific Ocean: A review of scientific results, 1985–1990. *Proceedings of the International TOGA Scientific Conference*, Honolulu, HI, July 16–20, 1990. Report WCRP-43, World Meteorological Organization, Geneva, Switzerland, 59–69.

McPhaden, M. J., and R. A. Knox, 1979. Equatorial Kelvin and inertio-gravity waves in zonal shear flow. *J. Phys. Oceanogr.* **9**, 263–277.

McPhaden, M. J., and B. A. Taft, 1984. Proceedings of the First International TOGA Workshop on Thermal Sampling, Rep. USTOGA-3, Univ. Corp. for Atmos. Res., Boulder, CO, 53 pp.

McPhaden, M. J., G. Reverdin, J. Merle, Y. du Penhoat, and A. Kartavtseff, 1984. Objective analysis of simulated equatorial Atlantic Ocean data on seasonal time scales. *Deep-Sea Res.* **31**, 551–569.

McPhaden, M. J., A. J.Shepherd, W. G. Large, and P. P. Niiler, 1991. A TOGA array of drifting thermistor chains in the western equatorial Pacific Ocean: October 1989–January 1990. NOAA Data Rep. ERL PMEL-34 (PB91-224535), 171 pp.

McPhaden, M. J., F. Bahr, Y. du Penhoat, E. Firing, S. P. Hayes, P. P. Niiler, P. L. Richardson, and J. M. Toole, 1992. The response of the western equatorial Pacific Ocean to westerly wind bursts during November 1989 to January 1990. *J. Geophys. Res.* **97**, 14,289–14,303.

McPhaden, M. J., A. J. Busalacchi, R. Cheney, J. R. Donguy, K. S. Gage, D. Halpern, M. Ji, P. Julian, G. Meyers, G. T. Mitchum, P. P. Niiler, J. Picaut, R. W. Reynolds, N. Smith, and K. Takeuchi, 1998. The Tropical Ocean–Global Atmosphere (TOGA) observing system: A decade of progress. *J. Geophys. Res.* **103**, 14,169–14,240.

Milburn, H. B., and P. D. McLain, 1986. ATLAS—A low cost satellite data telemetry mooring developed for NOAA's Climate Research Mission, paper presented at the Marine Data Systems International Symposium, Institute of Electrical and Electronics Engineers/Marine Technology Society, New Orleans, LA., 30 April–2 May 1986.

Miller, L., R. E. Cheney, and B. C. Douglas, 1988. Geosat altimeter observations of Kelvin waves and the 1986–87 El Niño. *Science* **239**, 52–54.

Moore, D. W., and S. G. H. Philander, 1977. Modeling the tropical ocean circulation. *The Sea*, Vol. 6, Interscience, New York, 319–361.

Namais, J., 1969. Seasonal interactions between the North Pacific and the atmosphere during the 1960s. *Mon. Weather. Rev.* **97**, 173–192.

National Research Council, 1983. El Niño and the Southern Oscillation: A scientific plan. National Academy of Sciences, Washington, DC, 72 pp.

National Research Council, 1990. TOGA: A review of progress and future opportunities. National Academy of Sciences, Washington, DC, 51 pp.

National Research Council, 1994. Ocean–atmosphere observations for short-term climate predictions. National Academy of Sciences, Washington, DC, 51 pp.

National Research Council, 1996. Learning to predict climate variations associated with El Niño and the Southern Oscillation. Accomplishments and legacies of the TOGA Program. National Academy Press, Washington, DC, 171 pp.

Nowlin, M., Jr., W. D. Briscoe, N. Smith, M. J. McPhaden, D. Roemmich, P. Chapman, and J. F. Grassle, 2001. Evolution of a sustained ocean observing system. *Bull. Am. Meteorol. Soc.* **82**, 1368–1376.

Nova University, 1978. The FGGE/INDEX/NORPAX Equatorial (FINE) Workshop. Nova University, Dania, FL.

Nova University, 1989. U.S. TOGA Observing System Mid-Life Progress Review and Recommendations for Continuation. Ft. Lauderdale, FL, 102 pp.

Obasi, G. O. P., 1998. Editorial in World Climate News, No. 13, World Meteorological Organization, Geneva, Switzerland, p. 3.

OOSDP, 1995. Scientific design for the common module of the Global Ocean Observing System and the Global Climate Observing System: An Ocean Observing System for Climate. Department of Oceanography, Texas A&M University, College Station, 265 pp.

Philander, S. G. H.;, 1990. *El Niño, La Niña, and the Southern Oscillation.* Academic Press, San Diego, CA, 293 pp.

Philander, S. G. H., and A. D. Seigel, 1985. Simulation of the El Niño of 1982–83. In: *Coupled Ocean–Atmosphere Models*, Elsevier Oceanography Series, Vol. 40, J. C. J. Nihoul, ed., Elsevier, New York, 517–542.

Picaut, J., E. Hackert, A. J. Busalacchi, R. Murtugudde, and G. S. E. Lagerloef, 2002. Mechanisms of the 1997–98 El Niño–La Niña, as inferred from space-based observations. *J. Geophys. Res.* **107**(C5), 10.1029/2001JC000850.

Rasmusson, E. M., and T. H. Carpenter, 1982. Variations in tropical sea surface temperature and surface wind fields associated with the Southern Oscillation/El Niño. *Mon. Weather Rev.* **110**, 354–384.

Reverdin, G., and M. J. McPhaden, 1986. Near surface current and temperature variability observed in the equatorial Atlantic from drifting buoys. *J. Geophys. Res.* **91**, 6569–6581.

Schopf, P. S., and M. J. Suarez, 1988. Vacillations in a coupled ocean–atmosphere model. *J. Atmos. Sci.* **45**, 549–566.

Servain, J., A. J. Busalacchi, M. J. McPhaden, A. D. Moura, G. Reverdin, M. Vianna, and S. E. Zebiak, 1998. A Pilot Research Moored Array in the Tropical Atlantic (PIRATA). *Bull. Am. Meteorol. Soc.* **79**, 2019–2031.

Stockdale, T. N., A.J. Busalacchi, D. E. Harrison, and R. Seager, 1998. Ocean modeling for ENSO. *J. Geophys. Res.* **103**, 14,325–14,355.

Taft, B. A., B. M. Hickey, C. Wunsch, and D. J. Baker, 1974. Equatorial Undercurrent and deeper flows in the central Pacific. *Deep-Sea Res.* **21**, 403–430.

Toole, J. M., and M. D. Borges, 1984. Observations of horizontal velocities and vertical displacements in the equatorial Pacific Ocean associated with the early stages of the 1982/83 El Niño. *J. Phys. Oceanogr.* **14**, 948–959.

Trenberth, K. E., and G. W. Branstator, 1992. Issues in establishing causes of the 1988 drought over North America. *J. Climate* **5**, 159–172.

U.S. TOGA Project Office, 1988. Report of a drafting workshop on elements of a ten-year monitoring component for El Niño and the Southern Oscillation (ENSO), May 11–18, 1983, AOML, Miami, Rep. USTOGA-1, Univ. Corp. for Atmos. Res., Boulder, CO, 42 pp.

van Oldenborgh, G. J., S. Y. Philip, and M. Collins, 2005. EI Niño in a changing climate: a multi-model study. *Ocean Science* **1**, 81–95.

WCRP, 1985. Scientific plan for the Tropical Ocean and Global Atmosphere Program, Tech. Doc. WMO/TD-64, World Meteorological Organization, Geneva, Switzerland, 146 pp.

WCRP, 1991. JSC/CCCO TOGA Scientific Steering Group, Report of the Tenth Session, Gmunden, Austria, 26–29 August 1991, Tech. Doc. WCRP-63, World Meteorological Organization, Geneva, Switzerland, 20 pp.

Wunsch, C., and A. E. Gill, 1976. Observations of equatorially trapped waves in sea level observations. *Deep-Sea Res.* **23**, 371–391.

Wyrtki, K., 1975. El Niño—The dynamic response of the equatorial Pacific Ocean to atmospheric forcing. *J. Phys. Oceanogr.* **5**, 572–584.

Wyrtki, K., 1985. Water displacements in the Pacific and the genesis of El Niño cycles. *J. Geophys. Res.* **90**, 7129–7132.

Wyrtki, K., E. Firing, D. Halpern, R. Knox, G. J. McNally, W. C. Patzert, E. D. Stroup, B. A. Taft, and R. Williams, 1981. The Hawaii–Tahiti Shuttle Experiment. *Science* **211**, 22–28.

Zebiak, S., 1989. Oceanic heat content variability and El Niño cycles. *J. Phys. Oceanogr.* **19**, 475–486.

Chapter 7

Reflections of an Equatorial Oceanographer

DENNIS WILSON MOORE

My career as an oceanographer began in the summer of 1958, after I finished my first undergraduate year at Harvard. I was a math major, but had managed to line up a job with Brackett Hersey at the Woods Hole Oceanographic Institution (WHOI), working on sound transmission data. I spent the summer playing data tapes on a big Ampex tape recorder, driving an analog output device that drew lines on a strip chart recorder. The strip charts were analyzed by using a planimeter to measure the area under the curve, which was proportional to the amount of sound energy received by a hydrophone used to collect the original data.

At the end of the summer I left for Bangkok, where my father was to spend the year teaching at Chulalongkorn University. I was on leave from Harvard, and spent the year teaching English. This included teaching at the Hydrographic Office of the Royal Thai Navy, where they were preparing for the arrival of oceanographers from Scripps to conduct the NAGA expedition. During the year the Office of Naval Research (ONR) liaison from WHOI came through Bangkok, and told me about a WHOI cruise that was taking place in the Mediterranean Sea the following summer.

Dennis Wilson Moore, Leader of the Ocean Climate Research Division, Pacific Marine Environmental Laboratory.

I sent a telegram to WHOI and asked to join the cruise. They said OK, and offered me \$250 per month, which was what I had earned the summer before. So I left Bangkok in May 1959 and made my way to Athens to join the R/V *Chain* cruise number seven. I got on the *Chain* in early June and got off in Woods Hole in mid-August. Memorable stops along the way included La Spezia, Monaco, Barcelona, and Bermuda. While we were at sea I stood watch in the top lab 8 hours a day, mostly operating the echo sounder and occasionally helping record thermistor chain data. Earl Hays was on the cruise with a new instrument called a sound velocimeter, built by the National Bureau of Standards. I helped him with the sound velocity profiling work and the analysis of the data, which was all done by hand. The WHOI oceanographers had recently realized that by putting a pinger on the bottom of a lowered instrument, and monitoring the profiling from the echo sounder, they could tell how far off the bottom the instrument was. This technique was developed and refined on this cruise. After the *Chain* returned to Woods Hole I lived aboard for another couple of weeks, until it was time to return to Harvard.

I continued my study of mathematics but I was already thinking seriously about being an oceanographer. I applied for a summer fellowship at WHOI for the following summer, and was fortunate to receive one. I was assigned to work with Claes Rooth, and we both sat in Al Woodcock's lab on the second floor of the Bigelow Laboratory. Andy Ingersoll was there as well. He and Woodcock had been doing laboratory studies on the bursting of air bubbles at the sea surface, and the resulting jet formation that expelled small water droplets into the atmosphere. Claes and I were supposed to find a theory for that process, but made little progress. The main reason was that Woodcock's lab was directly across the hall from the small lecture room where the Geophysical Fluid Dynamics (GFD) course was being held for the second time. There were a couple of empty spaces in the room, and Claes arranged for us to occupy them as informal auditors. Louis Howard was the principal lecturer, and other lectures were given by Wilhelm Malkus, Henry (Hank) Stommel, Melvin Stern, and Joanne Malkus . The students included Dick Lindzen, Joe Pedlosky , Bill Holland, and Bob Blandford, among others. For me it was a challenging summer, because I was still an undergraduate and had not yet studied partial differential equations. I learned a lot in a hurry, but not in a very systematic way. Much of what I was exposed to only made sense some years later. A lot of the course focused on thermal convection, and Henry Stommel produced the Great Seal of the GFD course, featuring the dragon with a fire under its tail and a block of ice on its head.

By the end of the summer I was determined to be a theoretical oceanographer. I returned to Harvard for my senior year as a math major, and applied for graduate school. I applied to Harvard in Applied Mathematics, and Scripps in Oceanography. Henry Stommel was on the Harvard faculty at the time, and when I asked his advice he said, "Why don't you stay at Harvard and learn all the applied math George Carrier and his colleagues can teach you?" So I decided to stay at Harvard.

I started graduate school in the fall of 1961 and originally worked with Henry Stommel. One memorable project was a study that I did while taking his oceanography course the following spring. He was interested in the dynamics of point vortices in a circular ocean basin on a beta plane, and had arranged with Ed Fredkin to help with the computations. Fredkin had a programming business in Maynard, Massachusetts, in the same warehouse with the Digital Equipment Corporation, and he would occasionally get a block of time on one of their machines. The only problem was, it was always midnight until 8 AM, and we did not find out about it until late the night before. Hank would call me around 10 PM, I would drive to Lexington to meet him, and we would proceed to Maynard for a night of computing. We had a lot of fun with this problem. I was able to show that vortices on a beta plane should accelerate near a western boundary and decelerate near an eastern boundary.

In the summer of 1962 I sat in Henry Stommel's lab at WHOI, along with Tommy Rossby, who was doing thermal convection experiments using mercury as the working fluid. Eric Kraus sat across the hall and we often played chess at lunchtime. Early in the summer I saw a notice about a summer school on air–sea interaction that Henry Charnock was organizing at Imperial College, London. I wrote Charnock but got no reply. I mentioned this to Kraus, who suggested I go talk to Columbus Iselin, the former WHOI Director. Iselin said he thought I should go to the meeting. He picked up the phone and told Henry Behrens, the WHOI comptroller, to arrange it! Richard Goody and Allan Robinson, who ran the program I was enrolled in at Harvard, were surprised to see me in London. They had brought Stan Jacobs and Victor Barcilon to the meeting, and did not expect any other Harvard students. I had a great two weeks in London and learned some basics of air–sea interaction and ocean circulation. During the 1962–1963 school year I completed my graduate course requirements.

In 1963 Henry Stommel decided to move to MIT, and I decided to stay at Harvard. I felt comfortable there, and the Committee on Applied Mathematics only required reading knowledge of one foreign language. I already knew enough German to pass that requirement. So I switched to Allan Robinson as my adviser. I spent a fair amount of time working on the computation of second-order nonlinear effects for barotropic Rossby waves on a midlatitude beta plane. I stayed in Cambridge in the summer of 1964 but Allan Robinson and Francis Bretherton both were at the GFD course at WHOI. Bretherton gave a lecture on time-dependent motions near the equator, and Robinson suggested to me that I investigate the baroclinic waves on an equatorial beta plane, with a view toward studying the nonlinear effects. So starting in the fall of 1964 I began to work on equatorial waves. Rattray (1964), Blandford (1966), Matsuno (1966), and Rattray and Charnell (1966) all published important papers on the topic. Rattray and Charnell tried to find equatorial wave solutions in a closed basin, and I was not satisfied with the approximations they used to obtain their result. The focus of my thesis turned out to be how to construct wave solutions in an equatorial basin bounded by eastern and western boundaries, but open to the north and south. The modes I constructed all had energy propagating equatorward from high latitudes on the western boundary and an equal energy flux going poleward on the eastern boundary. Michael Longuet-Higgins was producing similar modes on a sphere at the same time. For some reason I never looked at reflections off an eastern or a western boundary separately, but only studied basins with boundaries on both sides.

My fellow graduate students included George Philander and Conrad Lautenbacher (the present NOAA Administrator), among others. For the story of Philander's first meeting with Jule Charney, see the chapter by Stommel and Moore on Charney's contributions to oceanography in *The Atmosphere—A Challenge: The Science of Jule Gregory Charney*, edited by Richard S. Lindzen, and published by the AMS in 1990.

Postdoctoral fellows working with Allan Robinson while I was a graduate student included D. James Baker (who was NOAA Administrator under Bill Clinton) and Pearn P. Niiler. During the 1966–1967 school year Robinson was on sabbatical in India, and Peter Stone acted as my adviser in his absence. By then I had a pretty clear idea of what I was trying to do, and Peter was very supportive. By early 1967 Niiler had left Harvard to join the faculty at the Oceanographic Laboratory of Nova University in Fort Lauderdale, under the direction of William S. Richardson. Niiler called me one day and asked if I would like to visit Nova, with a view toward coming there after finishing my degree. I visited in May 1967. At that time the faculty in oceanography consisted of Bill Richardson, Peter Niiler, and Bill Schmitz. I also visited the University of Washington at the invitation of Maurice Rattray. I eventually accepted an offer to come to Nova. I handed in a draft of my thesis in the fall of 1967, and defended it in March 1968. I went to Scripps for six months for a short postdoctoral stay with Walter Munk, and then I moved to Nova in the fall of 1968. By the time I arrived Dayton Carritt, Charlie Yentsch, and Russ Snyder had also joined the faculty.

Bill Richardson had a unique approach to studying the ocean. At this time oceanographic and atmospheric studies were carried out in three distinct modes. Direct observations of the atmosphere and oceans were essential, theoretical (and eventually numerical) modeling studies were conducted to make sense of the observations, and laboratory scale experiments were conducted to allow the exploration of the parameter space associated with the phenomenon of interest. This was the era of large-scale laboratory experimentation, using elaborate rotating turntables. Richardson had been at WHOI and the University of Miami before he set up the lab at Nova, and he made a conscious decision to concentrate on direct observations of the ocean, along with enough theory to make sense of the observations. There was a specific decision not to get involved in laboratory experiments, but to concentrate on novel ways of measuring how the ocean works. Richardson and Schmitz were concentrating their efforts on measuring the velocity structure and transport of the Florida Current, at seven sections from the Florida Keys up to Cape Fear (see Niiler, 1975). The key to the measurement technique was a navigation system called Decca Hi-Fix, that had accuracy comparable to today's Global Positioning System (GPS). This allowed Richardson to develop a dropsonde instrument that included a free fall STD and vertically integrated transport measurement. Other instruments that drifted on the surface or dropped weights at intermediate depths allowed a current profile to be measured. Further refinements on the basic idea of these instruments led to the White Horse profiler that Schmitz used off New England to measure the Gulf Stream, the profiler that Luyten and Swallow used in the Indian Ocean in the mid-1970s to first describe equatorial deep jets, and the profiling work that Eric Firing did across the equatorial undercurrent south of Fanning Island during PEQUOD in 1982–1983. The development of the instruments that enable us to learn how the ocean works has been accomplished through a gradual evolution that has depended on significant engineering efforts to bring new technology to bear on the problem of making accurate *in situ* measurements. Without the *in situ* measurement capability everything else that we do would be for naught.

I got to Nova in the fall of 1968 and worked closely with Peter Niiler. He had visited me briefly while I was at Scripps and suggested that we look at the Jule Charney model for an inertial Gulf Stream to see if we could extend it beyond the separation latitude. We started in on this problem when I got to Nova and pursued it at the National Center for Atmospheric Research (NCAR) during the summer of 1969. Elliott Schulman and Jim O'Brien were there, and Elliott had arranged for a number of oceanographers to visit and lecture that summer. Adrian Gill and Allan Robinson were among them, and Henry Stommel was supposed to come, but had to bow out due to a broken rib. We watched the first landing on the moon with a large group in Jim O'Brien's living room. Jim and I sat in the back, playing cards. The game is called Klabberjass, or Klob.

In June 1970 I attended the Congress on Applied Mechanics in Cambridge. Bill Richardson asked me to represent Nova at a meeting that Allan Robinson was convening to discuss a program to study the role of eddies in the deep ocean circulation. I conveyed Richardson's interest in taking part in the program, with a view toward making current measurements from aircraft. I did not think much more about this until the following May, when I got a call at Nova from Henry Stommel and Allan Robinson, who asked if I would consider coming back to Cambridge for 2 years to work at MIT as the Executive Officer for the Mid-Ocean Dynamics Experiment (MODE). This was the program to study the role of eddies in the ocean circulation, and was jointly funded by the National Science Foundation (NSF) and ONR as part of the International Decade of Ocean Exploration (IDOE). Stommel and Robinson were the cochairmen of the MODE program. I visited the program managers in Washington, D.C. (Worth Nowlin at NSF and Denny Kirwan at ONR), and then told Stommel and Robinson I would do it. Bill Richardson was not too happy about this, as he was worried about the impact on my research. He eventually wrote a play on the subject, called "Mid-Ocean Madness, or I'm in the MODE for Love." I never knew for sure why Stommel and Robinson asked me to take on this job. Robinson told me that he and Hank were in his office, and Allan was looking at the collection of Ph.D. theses that he had supervised. Apparently he thought I might be able to do it.

Robinson was due to be on sabbatical in England and India that year, so we agreed to put the MODE office at MIT next to Henry Stommel in the Green Building. Norm Phillips was the department chairman and Jule Charney was just down the hall. Carl Wunsch was one floor down. The MODE program involved 20 principal investigators from 12 U.S. institutions and the National Institute of Oceanography (NIO) in England. There was an extensive management structure, with two cochairmen and an executive officer (a.k.a. the "troika"), an executive committee, a theoretical panel, and at least six other committees. The 4-month-long field experiment was to take place south of Bermuda during March–June 1973, and in the meantime everyone was busy building instruments, designing arrays, making preliminary measurements, and attending planning meetings. The principal investigators (PIs) constituted a group called the MODE Scientific Council, which met every 2 months beginning in July 1971. There were many new instruments developed for MODE, including the WHOI Vector Averaging Current Meter, the SOFAR float, a bottom-mounted pressure and temperature recorder, the electromagnetic current profiler, and expendable current probes launched from aircraft.

My job was to keep track of everything that was happening without telling the PIs how to run their projects. I talked with everyone about how things were going, planned the Scientific Council meetings, produced the needed documents, and tried to be sure we were on schedule. I had lots of help from Hank, Allan, and the rest of the Executive Committee. Everyone wanted the program to succeed.

A great deal of time was spent figuring out how to design the observing array. During the summer of 1972 the theoreticians all went to NCAR for 10 weeks to work out an experimental design, while the observationalists of the "Array Committee" held a 2-week session at WHOI to put together their version of what should be done. Russ Davis was the only person to attend both, so his ideas were well represented. The MODE Executive Committee met weekly, often by conference phone. The meetings were noteworthy for a number of reasons, including the number and strength of the cigars that were smoked, as well as the loudness of some of the participants. Students in the Green Building maintained that when we met on the 14th floor we could be heard clearly on the 13th and 15th floors as well. Francis Bretherton and I were the loudest! (See Chapter two in this volume.)

In the fall of 1972 there was a meeting on Numerical Models of Ocean Circulation sponsored by the National Academy of Sciences (NAS) and held in Durham, New Hampshire. The meeting organizers felt it was important for the numerical modelers to know something about the measurements that would be available to test their model results against. Nick Fofonoff and I wrote a paper on modern measurement techniques, which I presented (see Fofonoff and Moore, 1975). The Proceedings volume from the meeting was published in 1975. Henry Stommel's remarks in the conclusions section are well worth looking up (see Munk *et al.*, 1975). He was talking about the different styles of doing science that were in vogue at the time. IDOE was in full swing, including MODE, NORPAX and CUE, and GEOSECS, all more or less highly organized and involving lots of players. There were also the individual investigators, like Rattray and Veronis. Stommel went on, "I see the motorcycle gang, too—the Niilers and Gills and Welanders of this world. They are on their own, but I don't know how long they are going to last.... Then there are some very big buses, with not very many people on them, and they seem to have UNESCO diplomatic plates, and everybody's scared they are going to use up all the gas. Then I see one lonely individual, Arthur D. Little, who's been hired by the National Data Buoy Project, who is handing out questionnaires like, 'Do you like your speedometer to be calibrated in miles per hour or kilometers per hour?' " Some present-day concerns are not new.

The MODE experiment was conducted successfully and on schedule, and involved six ships, two aircraft, our own radio station and data center on Bermuda, a private phone system linking the Bermuda data center to Harvard, MIT, Yale, WHOI, and Johns Hopkins, and a network of Xerox telecopiers that could transmit a page in either 4 or 6 minutes. These were used to share data in near real time. There was plenty of excitement. Eli Katz aboard the *Chain* forgot about the watch circle around each subsurface mooring location, and steamed directly over one of Bob Heinmiller's moorings with his towed CTD instrument. He snagged the mooring with the towfish, and could not get loose. So he called me at the command post in Bermuda and asked for the release codes for Heinmiller's acoustic release, so he could pop the mooring and recover his towfish. Needless to say, Bob was not amused.

The main messages I came away from MODE with were that big science could be a lot of fun, that it was important to invest resources in the development of new instruments and techniques, and that while there was a significant role for theoreticians in planning these experiments and interpreting the results, the people who actually made the measurements needed to play the primary role. The other thing that I got from the 2 years at MIT were a set of lasting friendships and collegial relationships that substantially shaped the rest of my career. Ed Sarachik came to MIT to work with Jule Charney, and sat in the office next to mine. Mark Cane, Antonio Moura, and Inez Fung were all graduate students there at the time. Charney was working with George Philander and others to try to design the GARP Atlantic Tropical Experiment (GATE), to be carried out in the summer of 1974. Charney asked Sarachik to build a theoretical model of the tropical ocean, including the mixed layer and the undercurrent, to help guide the design of GATE. Sarachik got hold of Lighthill's 1969 paper on "Dynamic Response of the Indian Ocean to the Onset of the Southwest Monsoon," and as he read it he started asking me questions about equatorial waves. He forced me to consider the effect of a boundary on one side of the ocean basin or the other. This led me to realize the fundamental difference between reflection of baroclinic equatorial waves at western boundaries and at eastern boundaries. At a western boundary an incoming wave of a given meridional mode number n can always be reflected by a finite set of outgoing waves at mode number n or lower, plus a Kelvin or a mixed Rossby–gravity (a.k.a. Yanai) wave. But at an eastern boundary the same is not true, because equatorial Kelvin and Yanai waves only carry energy eastward. Therefore, reflections at an eastern boundary involve higher meridional mode numbers, and eventually include modes decaying to the west which result in energy propagating poleward along the boundary at high latitudes.

The Lighthill paper had left out the Kelvin wave completely and as a result had not gotten the western boundary response correct. The model that Sarachik started to develop was eventually handed off to Mark Cane to complete. He and Sarachik wrote a comprehensive series of papers on barotropic and baroclinic disturbances, with special attention to the equatorial wave guide.

Before I left MIT Henry Stommel asked me if I could assemble a group of equatorial theoreticians to think about possible experiments in the Indian Ocean. I got Dick Lindzen, George Philander, and Ed Sarachik to agree to serve on the INDEX Theoretical Panel.

I returned from MIT to Nova in the fall of 1973. George Philander contacted me and asked if I could come to Paris for a meeting at IOC to help plan GATE oceanography. I stopped in Boston on the way and went to Woods Hole to see Henry Stommel. Michelle Fieux was there visiting as well, and one evening she and Stommel used a sextant to measure the angular separation between two stellar objects while I was given the ephemeris and nautical almanac from which to calculate the theoretical answer. The results agreed to within the error of the sextant observation, much to our pleasure. I also met with Mark Cane to get his ideas for GATE, because he could not go to Paris. The airport at Boston was full of troops when I left for Paris, as a war had just broken out between Israel and the Arabs. The meeting in Paris was great fun. The participants were an interesting, enthusiatic group. They included John Knauss and a couple of his students, Walter Duing from Miami, Philip Rowlands, and a significant contingent of both East and West Germans, as well as a few Russians. One morning I opened the *Herald-Tribune* to see that U.S. Vice President Spiro Agnew had resigned. I told Knauss there was a job for him in Washington, but he was not interested. He later became the NOAA Administrator under George H.W. Bush.

In terms of planning for some oceanography at the equator, we decided GATE should concentrate on the undercurrent, which was the only significant feature we knew anything about. The meteorologists primarily wanted the oceanographic vessels as observational platforms for the atmosphere, and had no strong ideas about what oceanography we should do. We laid out an observational plan that involved a zigzag pattern as the ships steamed along the equator, to allow monitoring the position of the undercurrent and lowering current profilers through it. By the time the experiment was actually run, the transects were done in the opposite direction from what we had suggested, and the mean core of the undercurrent was observed to be some distance off the equator. There were also significant fluctuations of a biweekly time scale (see Duing *et al.*, 1975).

The year I came back from MIT to Nova, Niiler and I wrote up our Gulf Stream separation model and submitted it to George Veronis for the *Journal of Marine Research*. He accepted it for publication the day we submitted it. The work involved a complicated sequence of nested boundary layers. I was rather proud of it, but Carl Wunsch later cited it as an example of "analytic models . . . which . . . seemed untestable and intuitively implausible outside the laboratory." I still like to work on problems of this sort (see Moore and Niiler, 1974, and Wunsch, 1981).

In the spring of 1974 Henry Stommel came to Miami to meet with Walter Duing, Ants Leetmaa, and me, to plan an Indian Ocean program (INDEX) to propose to NSF and ONR. We all spent an afternoon on Walter Duing's sailboat in Biscayne Bay, laying out a possible program. We got our colleagues to write the necessary proposals, and we wrote a brief cover document for the program. The Global Atmospheric Research Program (GARP) had already announced that the First GARP Global Experiment (FGGE) would take place in 1977 and 1978. Our idea was to do pilot studies to allow us to plan a successful oceanographic component for FGGE in the Indian Ocean.

Later that spring, Stommel and I attended a meeting that Joe Smagorinsky hosted at GFDL to start discussing a monsoon program for the FGGE. We laid out our plans for INDEX. Joe Smagorinsky asked how well we needed to know the wind field that was driving the system. I said that in the tropics we needed daily averages on 1-degree squares, unless someone could demonstrate that the winds were coherent on longer scales. Smagorinsky said it would be a long time before we got that sort of resolution, but that some engineers at NASA thought it could eventually be done from satellites. Only now has this become possible.

In the summer of 1974 I went to WHOI to take part in the GFD course. I stopped at GFDL in Princeton on my way to Woods Hole, to start work with George Philander on our chapter for volume 6 of *The Sea* (see Moore and Philander, 1977). Early in July, Henry Stommel and I made a day trip from Woods Hole to NSF in Washington, D.C. and delivered the first set of INDEX proposals to Gene Bierly. These were eventually funded, and we set up an INDEX program office at Nova, with Jan Witte as the coordinator.

At the end of the summer as I was returning to Fort Lauderdale I met Bill Richardson in Washington, D.C. He was on his way to Maine with the Nova R/V *Gulfstream*. His plan was to spend the fall at the Bigelow Laboratory in Boothbay Harbor, Maine, testing a satellite-tracked drifting buoy he was developing for the NOAA National Data Buoy Center, for use in the southern oceans during FGGE. His approach was to launch buoys on day cruises out of Boothbay, and subsequently track their progress via research aircraft. Peter Niiler was about to move from Nova to Oregon State University, and Richardson asked me to watch out for the laboratory while he was away. By this time Charlie Yentsch had left Nova to take on the Directorship of the Bigelow Laboratory, Dayton Carritt had moved to the University of Massachusetts, and Mark Wimbush had joined the Nova faculty.

In early January 1975, Bill Richardson and four others sailed from Boothbay Harbor aboard the R/V *Gulfstream* and never returned. There was a severe storm that started the next day and the Coast Guard investigation concluded that the ship foundered during the storm. The body of one crew member was eventually found. Ironically, Richardson had just become the chairman of the University–National Oceanographic Laboratory System (UNOLS) Advisory Committee that month. His loss prompted UNOLS to promulgate a set of safety standards.

The loss of Richardson and four others put the lab at Nova in a terrible situation. Through the hard work of those left at the lab at Nova, and the generous support of the University and the community, the laboratory survived. I was made Acting Director, and eventually Director. Within a few months things began to settle down, and we got back to our research.

In May 1975 the INDEX Theoretical Panel held its first meeting. There were a dozen participants. There were not many people doing equatorial theory in those days, so the panel served as a training ground for young theoreticians. By the next year, there were plans underway for equatorial oceanography during FGGE in all three oceans, under the Scientific Committee on Oceanic Research (SCOR) Working Group (WG) 47, chaired by Stommel. He set up a panel for each ocean, chaired by Taft for the Pacific, Hisard for the Atlantic, and Swallow for the Indian Ocean. I had already entrained all the theoreticians to work on the Indian Ocean, so Taft and Hisard wrote and asked me to expand the scope of the panel to include the other two oceans. Thus, the INDEX Theoretical Panel became the Equatorial Theoretical Panel.

In December 1975 I attended a SCOR WG 47 Pacific Panel Meeting in Hawaii, at the Kuilima Resort on the north shore of Oahu. This was my first visit to Hawaii and I was fascinated with the place. Bruce Taft chaired the meeting. Klaus Wyrtki and Colin Ramage took care of the local arrangements. They put us in an out-of-the-way place with few distractions. I met a student from Scripps named Julian (Jay) McCreary, who was working on a thesis with Myrl Henderschott. It was a linear equatorial wave model for El Niño, and he was having trouble convincing his committee that he had solved the problem correctly. I went back to Florida after this meeting and wrote a letter to Russ Davis and Myrl, trying to convince them that Jay was doing good work.

In the spring of 1976 there was a UNESCO meeting in Nairobi on Cooperative Investigations of the North and Central Western Indian Ocean (CINCWIO). John Swallow, Henry Stommel, Walter Duing, Fritz Schott, and I were all there, along with many others. One of the participants was Mr. J. Findlater, who had discovered and described the low-level atmospheric jet trapped against the east African highlands during the Southwest monsoon. This is basically a western boundary current in the atmosphere, supported by a mountain range. In subsequent documents about INDEX we emphasized that it was timely as a contribution to CINCWIO. Later that spring Luyten and Swallow were at sea in the Indian Ocean and discovered the equatorial deep jets, which they referred to as "Equatorial Undercurrents." Carl Wunsch was with them, and produced a theory that turned out to be unsatisfactory. It is interesting to note that it is nearly 30 years later and there is still no satisfactory theory for these deep jets (see Luyten and Swallow, 1976, and Wunsch, 1977).

In the summer of 1977 we held the FGGE, INDEX, NORPAX, Equatorial (FINE) Workshop at Scripps, to complete preparations for the equatorial oceanography to be carried out in FGGE. This workshop began with 100 participants for the first week, with lectures by myself, Carl Wunsch, John Boyd, George Philander, Walter Duing, David Anderson, Julian McCreary, John Allen, and Ed Sarachik. The second week continued with about 24 participants, and a series of lectures by John Boyd. The afternoons were reserved for general scientific discussion. For the next 4 weeks a dozen participants stayed on and worked together on a variety of problems. This included myself and Jan Witte from Nova, Jim O'Brien and David Adamec from

FSU, Jay McCreary and Mike McPhaden from Scripps, and Phillipe Hisard, Jacques Merle, Joel Picaut, and Jean Marc Verstraete from France. At the end of the summer Jay McCreary defended his thesis, and I hired him to join the faculty at Nova. By the time he got there, the Office of Naval Research had already agreed to fund his work.

In February 1978 there was a NORPAX meeting at Mt. Hood. Jay and I attended from Nova. Klaus Wyrtki was there from Hawaii, and was the chairman of NORPAX by this time. One evening Klaus and I were sitting in the heated outdoor pool, and he asked me why I had not applied for the job that they had been advertising in Hawaii. This was the Joint Institute for Marine and Atmospheric Research (JIMAR) Director position. I did apply, paid a visit to the University of Hawaii (UH), and was eventually offered the job. I agreed to start in March 1979, with frequent visits in the interim.

Some of my most pleasant memories are associated with various educational undertakings. One of these was lecturing at a series of "Expert Lectures on Ocean Circulation Modelling" that Jim O'Brien organized at the Ecole Polytechnique in September 1978. I attended for a week, and was training for a marathon at the time. One morning I ran from the Ecole Militaire, next to my hotel, all the way to the Ecole Polytechnique, about 30 kilometers. It took a bit longer than I expected, and I was late to class.

Back at Nova, Jay McCreary had constructed a linear model for the equatorial undercurrent, and I encouraged him to talk about it at the fall AGU meeting in San Francisco. We got Jack Frost, the machinist and jack-of-all-trades at Nova, to make a plexiglass model showing the equatorial section and five perpendicular sections, as a three-dimensional visual aid to understanding the model. Jay's AGU talk was well received. Jay still has the plexiglass model. The paper (McCreary, 1981) was published in the *Philosophical Transactions of the Royal Society*.

I went directly from San Francisco to Hawaii, where I was scheduled to run a marathon. I spent an hour on the phone with Bob Harvey on Saturday morning, talking about plans for JIMAR. We broke off the conversation when he had to leave for a weekend cruise on the *Holo Holo*.

I ran my marathon on Sunday, slower than I had hoped. I went in to UH on Monday, expecting Harvey to appear later that day or early Tuesday. He was supposed to get off on Monday morning and return to Honolulu. The *Holo Holo* never made it in, and was not heard from or seen again. Ten people were lost at sea on this cruise, including the owner of the vessel, three UH scientists, two NOAA scientists from the Pacific Marine Environmental Laboratory (PMEL), and four others. The Coast Guard searched for more than a week. Because of my earlier experience with the loss of the *Gulfstream*, I was in the center of the UH effort to help with the search. Eventually one instrument box washed up on a beach on the big island. Nothing else was ever found. This tragic loss, occurring just 4 years after the *Gulfstream* disappearance, should never have happened. The Coast Guard concluded that the vessel was not sea-worthy. The University of Hawaii was (and still is) a member of UNOLS, but the people responsible for chartering the *Holo Holo* knew nothing about the safety standards that UNOLS adopted after the loss of the *Gulfstream*. Subsequently UNOLS adopted a separate set of standards and procedures for charter vessels. Unfortunately, all of this illustrates that oceanographers are no better than anyone else at learning from their mistakes. It often takes more than one.

In March 1979 I moved to Hawaii and became the JIMAR Director. The FGGE fieldwork was already underway, and we were busy planning what to do next. A group who wanted to have a detailed look at the equatorial waveguide proposed an experiment called PEQUOD, which stood for Pacific EQUatorial Ocean Dynamics. Jim Baker, then at the University of Washington (UW), and I were the cochairmen for this program, and in early June we put together a cover document to accompany the proposals. Shortly thereafter Jay McCreary arrived in Hawaii for a summer visit at JIMAR, and he and I immediately set about reworking the document. We had Jan Witte come out from Nova to help us. The cover documents were FEDEXed to NSF in time to be there by the July 1 deadline, only FEDEX lost the shipment. So we had to produce another set of copies the following week, and this time we shipped them to Baker in Seattle, to hand carry to Washington. The program was eventually funded, but Baker's project was not, so he dropped out. Jim Luyten at WHOI became the other cochairman, and Bob Heinmiller at MIT took on the coordination responsibilities. Bob soon realized that the time difference between Honolulu and Boston was going to make communication difficult, and he contracted with GTE to set up a network using GTE TELEMAIL. We all got terminals that the phone could be connected to, and we had the equivalent of email a decade before the Internet became available. Heinmiller eventually took the system private and set up OMNET.

In December 1979 Jay and I attended the IUGG meeting in Canberra. Afterward we traveled with Rana Fine through New Zealand, Samoa, and Fiji. Jay and I concentrated on planning for the meeting of SCOR WG 47, to review what had been learned during FGGE. He also explained the ideas about the delayed oscillator theory for ENSO that he and David Anderson had been working on (McCreary and Anderson, 1984). The SCOR WG 47 meeting was held in Venice in April 1981. Jay was convener, and wore a jacket and tie! It was a memorable meeting. Some of the equatorial theoreticians had been having a disagreement, and the key words were "gullible" versus "skeptical". The argument had to do with sea level variability in the Gulf of Guinea and whether the cause was more likely to be a linear response to zonal wind variations in the western equatorial Atlantic or a nonlinear response to meridional winds in the eastern equatorial Atlantic. At some point during the week in Venice the theoreticians sat down at an outdoor café to work out a truce, but when Fritz Schott asked if he could join the discussion he was refused! The SCOR WG47 meeting results were summarized in *Recent Progress in Equatorial Oceanography*, edited by McCreary, Moore, and Witte, which appeared in December 1981.

The PEQUOD program took place starting in March 1982 and extended into 1983. It turned out to coincide with a major El Niño whose timing did not follow the accepted norm. So at a meeting at Princeton in the fall of 1982 we had a temperature record from Fanning Island showing the highest SST ever recorded there, and yet no one would admit that an El Niño was underway. (See the Chapter six in the present volume.) In the next half decade many equatorial programs were carried out.

A TOGA workshop in Honolulu in August 1986 resulted in another volume called *Further Progress in Equatorial Oceanography* (January 1987), edited by Katz and Witte. It includes articles on INDEX, NORPAX/FGGE, EPOCS, TOGA, SEQUAL, PEQUOD, Tropic Heat, WEPOCS, and the Equatorial Theoretical Panel. By this time there was no shortage of equatorial oceanographers.

During the early 1980s the FGGE project office under Rex Fleming and Mike Hall was transformed into the Office of Global Programs under Mike Hall and Ken Mooney. Klaus Wyrtki had a variety of sea level projects ongoing at Hawaii at the time, and Rex Fleming and Mike Hall encouraged him to establish a single activity called the Hawaii Sea Level Center, which they agreed to fund. This has been an ongoing activity for more than 20 years now, first under Wyrtki, then Gary Mitchum, and now Mark Merrifield. The sea level data are used both for climate studies and for tsunami detection, with most of the gauges capable of rapid sampling and real-time data transmission. The network also provides ground truth for satellite altimetry.

In March 1987 I called Henry Stommel to ask him about long-term sea level rise, but he immediately changed the subject. He sent me two papers on a laboratory demonstration of the Coriolis force which had been published by Arthur Holly Compton. The first was a 1913 paper in *Science* (Vol. 37, pp. 803–806), written when Compton was still an undergraduate. I read the Compton papers and tried to answer Hank's questions about them. The first one had an error in the theory, with a missing factor of 2, but Stommel could not believe that Compton got it wrong, and wanted me to explain the discrepancy. He then sent me a sketch of Woods Hole and WHOI partially submerged and the Eel Pond overflowing its banks(Figure 7.1). This was his contribution to the discussion of sea level rise! He also sent me a copy of the fourth draft of a book he had been working on called *An Introduction to the Coriolis Force*, and he asked me if I would be willing to be a coauthor and help him finish the book. He had originally tried to do the whole thing without using vectors, because he wanted the material to be easily accessible, even to high school students. I agreed to help, but only if we in fact used vectors to help tell the story. Working with Hank Stommel on this book was a wonderful experience for me. It was very intense, and involved lots of writing in the early hours of the morning. I would typically go to bed early, thinking about some problem, and wake up about 3 AM with an inspiration about what to write. I got to think through the details of many elementary mechanical problems we used to illustrate various concepts. Things I first learned about in the GFD course in 1960 finally made sense to me. The book appeared in 1989, and it was not a great success. But helping to write it was fun (Stommel and Moore, 1989).

By 1990 the biogeochemists were becoming interested in the equatorial oceans, and planning equatorial cruises in the Pacific. The only problem was that while there were lots of equatorial physical oceanographers, there were not many biogeochemists who knew anything about equatorial dynamics. So in June 1991 Lew Rothstein ran

April 1, 1987

Dear Dennis

Thank you for the pleasant phone call last night. It was very nice to speak with you again. The above picture is based upon a general rise of sea-level accompanying global warming. As you can see George Woodwell will have difficulty getting out of the MBL.

Good luck with Compton

(Science 1913, vol 37 pp 803 ff.)

Henry

Figure 7.1

a month-long program on Equatorial Dynamics at the University of Rhode Island (URI), specifically to help train the biogeochemists. Jay McCreary, Jim O'Brien, Lew Rothstein, and I formed the core faculty for the workshop. Mike McPhaden was there the first and last week, and a number of other lecturers came for a few days at a time. There were about 20 students. We all lived together in three or four rental houses, and shared cooking and housekeeping responsibilities. Jim O'Brien and Lew Rothstein set up a computer visualization laboratory for the workshop, using a number of Silicon Graphics workstations. So the students were not only learning about equatorial theory, they were also running time-dependent computer models to visualize what they were studying. At the end of the workshop we had a 2-day meeting of the Equatorial Theoretical Panel at the URI Whispering Pines Conference Center on the W. Alton Jones campus. Henry Stommel came to the meeting and I clearly remember an early morning discussion with him the second day. He brought me a goose quill he had picked up on the grounds, and suggested I could use it to write with. I think that was the last time I saw him.

Sometime late in 1991 I got a call from Chris Mooers asking if I would be interested in being an editor of the *Journal of Physical Oceanography* (JPO). I spoke with Peter Gent and Eli Katz about it, and said I would be willing to give it a try. I quickly joined the American Meteorological Society (AMS) and sent my CV to the publications committee. I became an editor in early 1992 and continued until late 1995. Editing an AMS journal provided a nice balance to my UH responsibilities as JIMAR Director.

In 1994 Bruce Taft announced that he would retire the following year, and when his job as Leader of the Ocean Climate Research Division at NOAA's Pacific Marine Environmental Laboratory was advertised, I applied. I was offered the job and moved to Seattle in 1995. It has been exciting to work closely with a diverse group of scientists who make *in situ* observations relating to the role of the ocean in climate. I have now been here 10 years and have no plan to retire. I enjoy my job. Lots of things have changed in NOAA over the decade. Because of my 17 years directing JIMAR in Hawaii, I still pay close attention to the operation of the NOAA joint institutes, especially JIMAR and the Joint Institute for the Study of the Atmosphere and Oceans (JISAO). My impression is that the joint institutes have been an excellent mechanism to promote cooperation between NOAA and the universities, and have been the locus of lots of outstanding research over the last 40 years. But I don't think directing a joint institute is as much fun as it used to be.

What lessons can we learn from all of this? The first one is that *in situ* measurements are essential for understanding how the system works. Remote sensing, theory, and numerical modeling are all important, but *in situ* observations using the best available instruments are the sine qua non of understanding the ocean. This requires continual investment in new sensor and instrument development, incorporating state-of-the-art technology, to enable scientists to continue to advance their knowledge. This need to keep investing in new measurement techniques was considered obvious in the early 1970s, and was built in to IDOE programs like MODE. It is no longer as easy to get the necessary resources to maintain the engineering development that in turn enables scientists to advance their knowledge. And a comparable investment in information technology and data handling infrastructure is also required, especially for the global climate problem.

Theory and modeling are very important for putting the pieces together and understanding how it all works. Only in rare cases does the theory precede the *in situ* observations. In the late 1960s someone said to me, "You don't really think all those equatorial waves exist in the ocean, do you?" During the following decade Wunsch and Gill (1976) found the equatorially trapped inertia-gravity waves in the sea level data, and then Knox and Halpern (1982) found the Kelvin waves in sea level and current meter data. Now these waves are recognized as the fundamental building blocks for understanding the role of the ocean in the climate problem.

ACKNOWLEDGMENT

I wish to thank my daughter Megan for her help in proofreading this chapter.

REFERENCES

Blandford, R., 1966. Mixed gravity Rossby waves in the ocean. *Deep-Sea Res.* **13**, 941–961.

Compton, A. H., 1913. A laboratory method of demonstrating the Earth's rotation. *Science* **37**, 803–806.

Düing, W., *et al.*, 1975. Meanders and long waves in the equatorial Atlantic. *Nature* **257**, 280–284.

Fofonoff, N. D., and D. W. Moore, 1975. Modern measurement techniques. In: *Numerical Models of Ocean Circulation.* National Academy of Sciences, Washington, DC, 57–60.

Katz, E. J., and J. M. Witte, eds., 1987. Further Progress in Equatorial Oceanography; A Report of the U.S. TOGA Workshop on the Dynamics of the Equatorial Oceans, Honolulu, HI, August 11–15, 1986.

Knox, R., and D. Halpern, 1982. Long range Kelvin wave propagation of transport variations in Pacific Ocean equatorial currents. *J. Mar. Res.* **40**(Suppl.), 329–339.

Lighthill, M. J., 1969. Dynamic response of the Indian Ocean to the onset of the Southwest Monsoon. *Philos. Trans. R. Soc. London Ser. A* **265**, 45–92.

Lindzen, R. S., ed., 1990. *The Atmosphere—A Challenge: The Science of Jule Gregory Charney*. American Meteorological Society, Boston.

Luyten, J. R., and J. C. Swallow, 1976. Equatorial undercurrents. *Deep-Sea Res.* **23**, 1005–1007.

Matsuno, T., 1966. Quasi-geostrophic motions in the equatorial area. *J. Meteorol. Soc. Japan II* **44**, 25–43.

McCreary, J. P., 1981. A linear stratified ocean model of the equatorial undercurrent. *Philos. Trans. R. Soc. London Ser. A* **298**, 603–635.

McCreary, J. P., Jr., and D. L. T. Anderson, 1984. A simple model of El Niño and the Southern Oscillation. *Mon. Weather Rev.* **112**, 934–946.

McCreary, J. P., Jr., D. W. Moore, and J.M. Witte, eds., 1981. Recent Progress in Equatorial Oceanography; A Report of the Final Meeting of SCOR Working Group 47 in Venice, Italy, April 27–30, 1981. Nova University/N.Y.I.T. Press.

Moore, D. W., and P. P. Niiler, 1974. A two-layer model for the separation of inertial boundary currents. *J. Mar. Res.* **32**, 457–484.

Moore, D. W., and S. G. H. Philander, 1977. Modelling of the tropical oceanic circulation. In: *The Sea, Vol. 6*, E. D. Goldberg, I. N. McCave, J. J. O'Brien, and J. H. Steebe, eds., Wiley–Interscience, New York, 319–362.

Munk, A. W., H. Stommel, A. S. Sarkisyan, and A. R. Robinson, 1975. Where do we go from here? In: *Numerical Models of Ocean Circulation*. National Academy of Sciences, Washington, DC, 349–360.

Niiler, P. P., 1975. Variability in western boundary currents. In: *Numerical Models of Ocean Circulation*. National Academy of Sciences, Washington, DC, 216–236.

Rattray, M., Jr., 1964. Time-dependent motion in an ocean: A unified two-layer, beta-plane approximation. In: *Studies in Oceanography*, dedicated to Professor Hidaka in Commemoration of His Sixtieth Birthday, Tokyo, 1962, 19–29.

Rattray, M., Jr., and R. L. Charnell, 1966. Quasigeostrophic free oscillations in enclosed basins. *J. Mar. Res.* **24**, 82–103.

Stommel, H. M., and D. W. Moore, 1989. *An Introduction to the Coriolis Force*. Columbia University Press, New York.

Wunsch, C., 1977. Response of an equatorial ocean to a periodic monsoon. *J. Phys. Oceanogr.* **7**, 497–511.

Wunsch, C., 1981. Low-frequency variability of the sea. In: *Evolution of Physical Oceanography*, B. A. Warren and C. Wunsch, eds., MIT Press, Cambridge, MA, 342–374.

Wunsch, C., and A. E. Gill, 1976. Observations of equatorially trapped waves in Pacific sea level variations. *Deep-Sea Res.* **23**, 371–390.

Chapter 8

Ocean Acoustic Tomography

From a Stormy Start to an Uncertain Future

WALTER MUNK

To understand the ocean, its dynamics, its role in climate, weather and other ocean-atmosphere phenomena, one must observe it on a basin-wide scale with adequate time and space resolution.[1]

INTRODUCTION

This is the story of the marriage of two disciplines: physical oceanography and ocean acoustics. It was a marriage of convenience, but with mutual respect, even love. I have spent the first thirty happy years of my career without much contact with acoustics, so it was a late learning experience.

This is a personal account, with no claim of historical accuracy and completeness. I have relied largely on my increasingly leaky memory. But I have tried to

Walter Munk, Secretary of the Navy Chair, Oceanography.

capture some of the spirit underlying the developments and discoveries of the last half-century.

THE MESOSCALE REVOLUTION

Ocean acoustic monitoring was introduced in the 1970s in direct response to the demonstration by the Mid-Ocean Dynamics Experiment (MODE) that the kinetic energy of the ocean circulation is mostly associated with variability on a relatively small scale, the mesoscale (order 100 km). The general circulation (time average) of the major ocean gyres contains only a small fraction, perhaps 1%, of the kinetic energy.

The classical physical oceanographers cast their Nansen bottles and contoured dynamic heights, so that these would be available for computing geostrophic currents, which are then published on permanent charts. Oceanographic vessels operating singly in the expedition mode were unable to cope with the spatial and temporal sampling requirements imposed by the newly discovered mesoscale variability. Expedition ships were unwilling (those under sail were unable) to stand still; the tradition "never to occupy a station twice" prevailed for over a hundred years. Textbooks by Krümmel[2] (1911), Sverdrup[3] *et al.* (1942), and Defant[4] (1945, 1961) all reflected the picture of a steady-state ocean. I was brought up in this tradition.

Satellites came just in time to provide the required sampling strategy for the mesoscale variability: global sampling with good spatial and adequate temporal resolution. These are three quite separate considerations, equally fundamental to the subsequent development of physical oceanography. But the interior ocean is opaque to electromagnetic radiation, yet transparent to sound. In the mid-1970s Carl Wunsch and I started thinking in terms of an acoustic system to fill this gap (Figure 8.1).

EARLY ACOUSTIC RUMBLINGS

The SOFAR channel (Sound Fixing And Ranging) was discovered by Ewing and Worzel in 1944. Soundspeed *increases* with both temperature and pressure, thus maintaining a minimum in soundspeed (and attendant waveguide) between the warm surface waters and the high-pressure abyss. Ewing and Worzel worked out the expected characteristics of the sound propagation, and then set about to prove the theory. *Saluda* departed Woods Hole with a deep hydrophone hung over the side. A second ship dropped 4-pound charges at distances up to 900 nautical miles. In the words of Ewing and Worzel, "the end of the . . . transmission was so sharp that it was impossible for the most unskilled observer to miss it." They spoke even then of transmission over 10,000 miles.[5]

Two years later the Russian acoustician Leonid Brekhovskikh independently discovered the ocean acoustic waveguide (personal communication): "Some work had

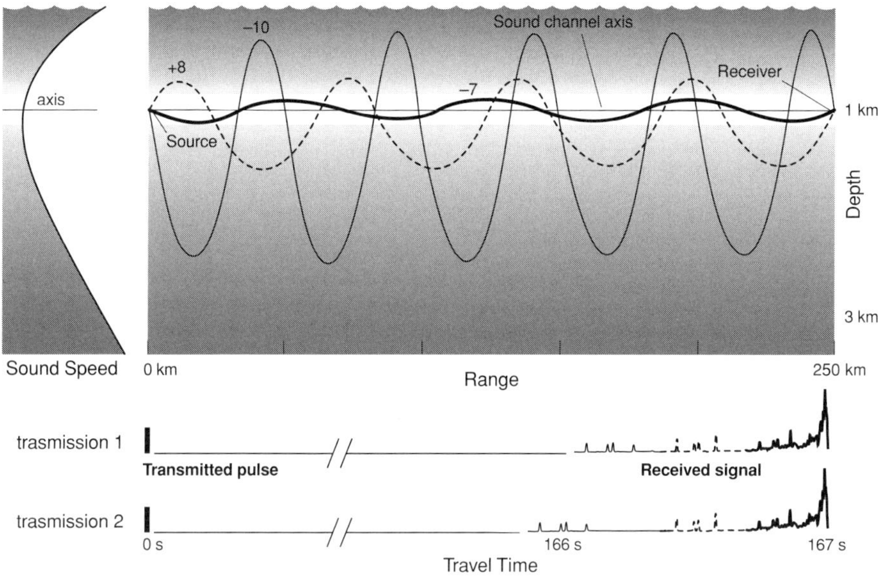

Figure 8.1. Cartoon of the Ocean Acoustic Tomography method. The minimum in soundspeed provides an axis for an acoustic wave guide. The time sequence of ray arrivals is from steep (thin) to moderate (dashed) to flat (heavy) launch angles (flat, axial rays cannot generally be resolved). A representative ray for each of the three classes is shown in the upper panel. Transmission 2 (relative to 1) shows earlier arrivals of steep rays, but no change otherwise. This is consistent with a warming of the upper ocean (or, less likely, a warming of the abyssal ocean), with intermediary and axial depths unchanged.

been planned for 1946 in the Sea of Japan, but the equipment was not ready. Rather than lose some ship time, it was decided to make some spontaneous measurements of sound transmission. Something very strange was observed; at large distances the signal started very weakly, then increased with time resembling a thunder in the final stage before coming to an abrupt end. It was my duty to treat these results. It appeared that the only way to explain them was to take into account the existence of an acoustic wave guide. . . ."

Thus, Ewing, the experimentalist, was confirming an existing theory of an ocean acoustic waveguide, whereas Brekhovskikh, the theorist, had stumbled on the discovery looking at data. I was a Scripps student visiting Woods Hole in winter 1944 and was enormously impressed with Ewing's orderly approach: from existing theory to subsequent experimental confirmation. (As it turned out, this was the last time in my career that I have known of a focused seagoing experiment to confirm a previously derived theory.)

A growing acoustic community under U.S. and USSR Navy sponsorships set out to exploit the sound channel for submarine detection. Brekhovskikh went on to write: "Since the work could have military applications its publication was delayed

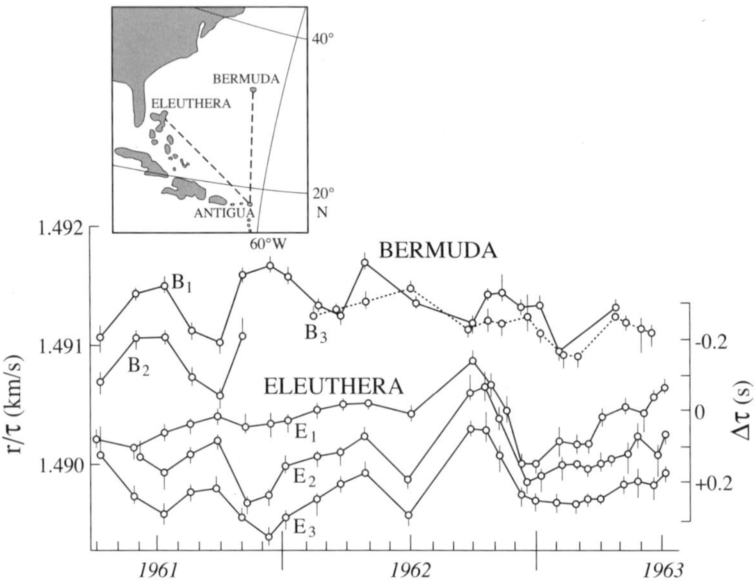

Figure 8.2. Hamilton's 1961–3 transmissions from explosive sources off Antigua to hydrophone receivers at Bermuda and Eleuthera.[6] Variations in travel time (plotted as acoustic group velocity) gave an early indication of mesoscale variability.

until 1948. . . . Due to the weakened international scientific relations we did not know of Ewing's paper until later."

Almost 20 years later, from 1961 to 1964, Gordon Hamilton conducted the Sound Channel Axis Experiment (SCAVE).[6] Precisely located and timed charges were fired off Antigua and received at Bermuda and Eleuthera at ranges exceeding 1000 km (Figure 8.2). Both source and receiver were near the sound axis. Over a period of 27 months, travel time variations up to 500 ms with time scales of several months were recorded. Hamilton remarks on the lack of correlation between the Antigua–Bermuda and Antigua–Eleuthera transmission (separated by 1000 km) as compared to the close correlation (but with significant differences) to the individual Eleuthera phones (60-km separation). This was perhaps the earliest clear indication of mesoscale variability. Unfortunately, the results were not published until 1977.

Hamilton's work was followed by Steinberg and Birdsall in 1966 with the pioneering MIMI transmissions (named for the Michigan–Miami collaboration) across the Straits of Florida.[7] In 1973 the Eleuthera–Bermuda transmissions were resumed by Clark and Kronengold. The essential new feature was that the explosive sources of earlier experiments had been replaced by piezoelectric transducers. These were 406-Hz CW transmissions with travel time perturbations deduced from the recorded changes in phase, a risky undertaking. We spent the next twenty years fighting for bandwidth to transmit pulse-equivalent signals.

Perhaps the most characteristic result of these early CW transmissions was the variability in receptions: fade-outs are the rule rather than the exception. Early attempts at understanding this inherent variability were in terms of a homogeneous and isotropic ocean fine structure. That assumption, though mathematically convenient, has nothing to do with reality: ocean fine structure is neither homogeneous nor isotropic. Garrett and I had developed a statistical model of internal waves, which was found consistent with the acoustic phase and intensity variances in the Florida and Eleuthera transmissions.[8] This work has been greatly extended by Dashen, Flatté, and others.[9] We now know that internal waves are the primary (but not the only) source of "noise" in the measurements of mesoscale induced travel time variations.

THE BEGINNINGS (1976–1980)

At a meeting in 1976 celebrating the thirtieth anniversary (1946–1976) of the Office of Naval Research, I described the transmissions between the R/V *Alexander Agassiz* and the R/V *Ellen B. Scripps* at 25-km range. These results were the basis for Worcester's thesis[10] (the first of many Tomography dissertations). There is a noteworthy asymmetry between the clean arrivals from beneath the sound axis and the complex multipaths through the upper ocean (Figure 8.3). The sum and difference

$$\tau_{\mathrm{sum}} = \tau_{\mathrm{A}\to\mathrm{B}} + \tau_{\mathrm{B}\to\mathrm{A}}, \qquad \tau_{\mathrm{dif}} = \tau_{\mathrm{A}\to\mathrm{B}} - \tau_{\mathrm{B}\to\mathrm{A}}$$

of the oppositely traveling transmission times yielded useful estimates of the deep and shallow temperatures and currents (sound travels faster in warm water and in the direction *with* the current).

By 1977 the first tomographic group was being formed. Spindel and Porter from Woods Hole brought their experience in underwater acoustics, autonomous instrumentation, setting deep-sea moorings, and precise acoustic position keeping. Webb had built acoustic sources for SOFAR floats. Birdsall and Metzger from the University of Michigan had developed the signal processing for the MIMI transmissions. Wunsch from MIT was pioneering the application of inverse theory to ocean observations. Worcester and Munk from Scripps participated. The formation of the group was something of a shotgun wedding under the persuasion of Hugo Bezdeck of ONR who wanted to avoid duplicate efforts; yet the group was to work together for a quarter-century.

In the fall of 1978 a 900-km transmission from a source southwest of Bermuda proved adequate to track 14 ray paths for the duration of the experiment.[11] This work came just in time to save tomography from an early demise. A reviewer of our proposal had written that individual ray arrivals could not be resolved, and even if resolved could not be identified, and even if identified would not be stable. The proposal was declined. We responded by sending Figure 8.4 with the sentence: "we have resolved, identified and tracked 13 stable arrivals for over 2 months (see figure)." The proposal was accepted.

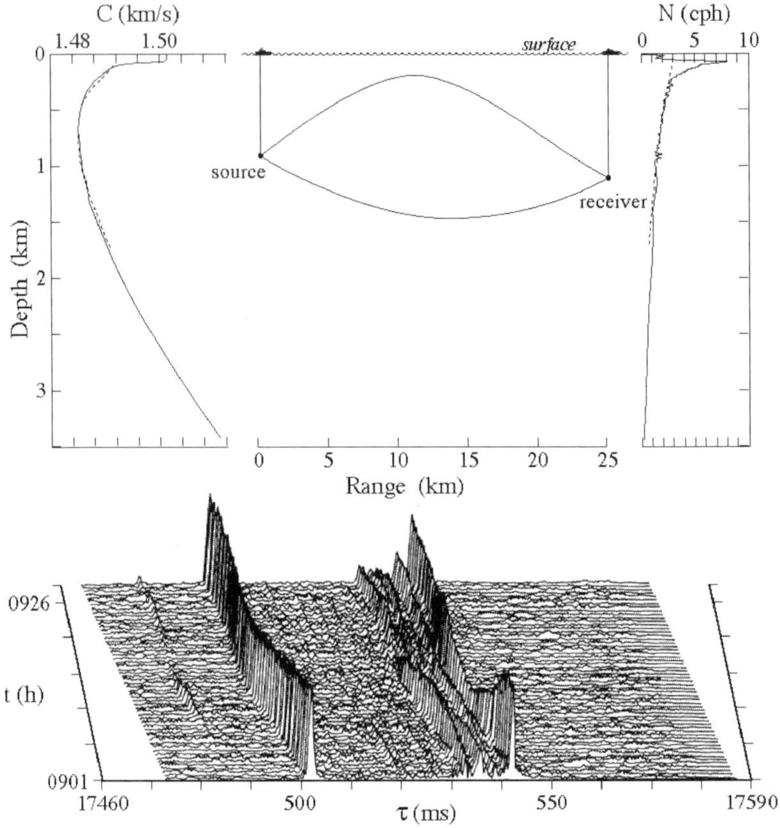

Figure 8.3. The first reciprocal transmission experiment was conducted by Worcester in 1976 off San Diego.[10] Top: The two ray paths for the 25-km range. Bottom: Intensity as a function of source time τ for transmissions at clock time t. Each transmission shows an early (deep) arrival in about 17.50 seconds followed by a late (shallow) arrival in 17.54 seconds. Successive transmissions during the 25-minute experiment show a complex and variable pattern for the upper path. Reciprocal transmissions (not shown) gave information about shallow and deep currents.

Carl Wunsch and I summarized the application of acoustic inverse techniques to monitoring an ocean basin for mesoscale fluctuations.[12] We concluded that such a system was achievable NOW. We chose the obscure name "ocean acoustic tomography" (in analogy with medical tomography for probing the interior of a brain by waves propagating through the brain in many directions) in the hope that our colleagues would want to find out what it was all about. The early emphasis was to produce contour charts with mesoscale resolution. The procedure going from transmission plots to the contour charts relies heavily on the application of inverse methods. There were (and still are) many loose ends; Carl and I spent a sabbatical year at the University of Cambridge to follow up some of the open topics.[13] Carl attended a Royal Society

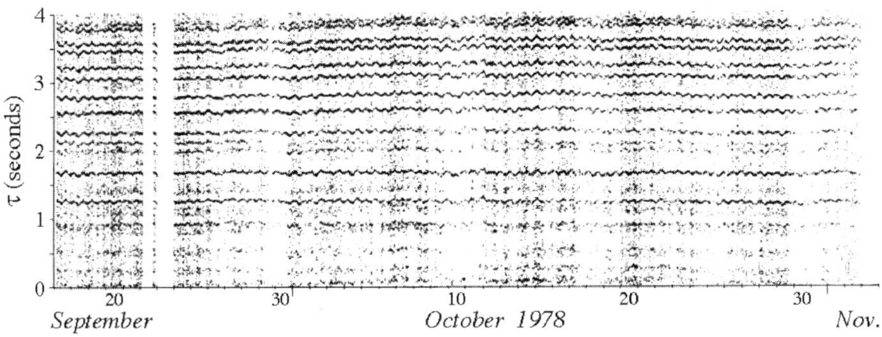

Figure 8.4. Arrival pattern at 900-km range from a moored Webb source (220 Hz) southwest of Bermuda.[11] Travel time τ (relative to the source pulse, and in relative units) is along the y-axis. Calendar time t is along the x-axis (reversed relative to Figure 8.3). Arrivals were recorded at 10-minute intervals over a 48-day period. A careful inspection shows a tidal component in the travel time.

Discussion Meeting concerning oceanography in the next decade. In "Observing the Ocean in the 1990's" we predicted a symbiosis of acoustic tomography and satellite altimetry[14] (but did not foresee the coming dominant role of profiling floats). Soon thereafter Ocean Acoustic Tomography made the title page of *Nature*.

THE ASW CONNECTION

Problems of classification and security are part of the oceanographic scene, especially in ocean acoustics with its ties to antisubmarine warfare (ASW). The years following Ewing's demonstration of the sound channel saw a rapid sequence of discoveries, including the anomalous absorption of sound in sea water, convergence-zone focusing, diurnal migration, and the biological origin of certain ambient noises. A growing acoustic community under U.S. Navy sponsorship set out to exploit the sound channel for submarine detection. Most of the work was classified, and when it was eventually reported in the open literature some 15 years later, the authorship bore little or no relation to the people who had done the pioneering work. In the meantime the acoustic and oceanographic communities had drifted apart, to the detriment of both.

In the development of acoustic tomography we tried to avoid splitting of the two communities. The development depended on collaboration with the Navy in two ways: for research support from ONR, and the use of Navy listening facilities. In 1978 westward transmission from Bermuda[11] was recorded at a "bottom receiver at $\approx$ 900-km range," referring to a Navy SOSUS array (*so*und *su*rveillance *s*ystem). The quotation is deliberately vague to honor an agreement to give ranges only to the nearest 100 km and travel times to the nearest 100 seconds (there is no restriction on *relative* travel times). The editor of the *Journal of Physical Oceanography* originally declined publication unless precise coordinates were published. We (successfully) defended the view that coordinates should be to the precision required for the scientific interpretation of the results. There is no simple answer here.

It is prudent to monitor source performance in real time, as has been demonstrated on a number of occasions. It is difficult to convey the satisfaction of hearing the signal arrive at a remote site at the expected moment. It is also difficult to convey the sense of disappointment when it does not.[15] Most of us have clearances to visit the facilities, and the monitoring in real time was arranged on a personal basis and initially did not involve our institutions. We are pleased to report that these arrangements (with some modifications) have worked satisfactorily. The tomographic community is greatly indebted to the U.S. Navy for having permitted the use of these remarkable facilities.

We are still, today, depending on the SOSUS arrays (as will be seen). But the situation has changed, and the Navy and the oceanographic community find themselves under a common siege by the environmental community.

A DECADE OF EXPERIMENTS (1981–1990)

The decade[16] started with the 1981 "Demonstration Experiment" (Figure 8.5a). Bruce Cornuelle, a Wunsch Ph.D., applied Gauss–Markov estimation theory to generate mesoscale maps for comparison with CTD surveys.[17] Most of the remaining decade was spent in a struggle to understand the capabilities and limitations of the method and to develop autonomous instrumentation adequate for megameter ranges. The 12 experiments, 8 in the Atlantic and 4 in the Pacific, are outlined in Table A.1 of MWW. Among the results were perturbations in travel time across the meandering Gulf Stream (Figure 8.5b)[18] and measurements of tidal currents and vorticity in a 1000-km triangle. The progression toward larger ranges culminated in the climate-oriented 4000-km transmissions by Spiesberger and Metzger from Hawaii to the mainland.[19]

I will single out the Greenland Sea Project 1988–89 by Worcester, Sutton, Morawitz, Cornuelle, Lynch, Pawlowicz, and others.[20] A 200-km pentagon array of transceivers was deployed in summer 1988, then frozen over, and recovered in summer 1989. The measurements yielded time histories of heat content, barotropic currents, and tides in an arctic environment and documented the formation and evolution of a convective chimney. The chimney had a spatial scale of about 50 km and a time scale of about 10 days, reaching depths of 1500 m.

Prior to the installation, concerns had been raised about the security of acoustic moorings in a strategic sea lane. (Moorings had mysteriously disappeared in the 1970 MODE experiment.) I telephoned Brekhovskikh in Moscow and arranged for a visit. My wife insisted on coming along (this was at the height of the Cold War). When we entered Brekhovskikh's office he held up his hand: "Let me tell you why you are here. You are here to make sure no one interferes with your moorings." I replied: "Yes." He said, "I will see what I can do, but I do not command the Soviet Navy." There was no interference. Some years later Brekhovskikh telephoned from Washington,

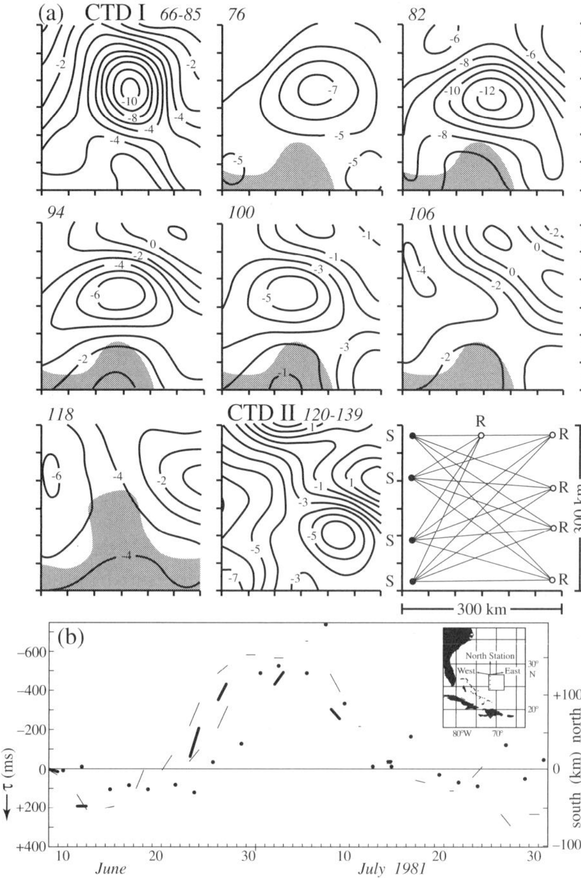

Figure 8.5. (a): Ocean "weather maps" from the 1981 Demonstration Experiment.[17] The panels show the soundspeed perturbations (m/s) at 6 day intervals at 700m depth, using Gauss-Markov estimates. Note the rapid change between 1981 year-day 100 and 106. For comparison, the 700 m temperature contours from CTD surveys at the start and end of the acoustic transmissions are shown. The experiment geometry of 4 sources and 5 receivers are in the 11th panel. (b): The "Experiment" featured a whopping 0.8 s change in travel-time associated with the meandering of the Gulf Stream.[18] Line segments denote intermittently measured travel time perturbations (left scale). Dots mark positions of the north wall of the Gulf Stream inferred from surface temperature measured by NOAA-6 satellite (right scale). As expected, travel times are short when the stream is furthest north, with the largest fraction of warm Sargasso Sea water along the path.

asking whether he could come to La Jolla for dinner (!). I had hoped to learn more about his role in the Greenland Sea experiment, but he showed up with an unexpected companion and the conversation never went beyond pleasant generalities.

Brekhovskikh's contributions to ocean acoustics have been a major factor in the history of ocean acoustic tomography. I last saw Brekhovskikh in Berlin at a meeting of the European Geophysical Society where he received the Munk Award given jointly by the U.S. Navy and The Oceanography Society.

HEARD ISLAND FEASIBILITY TEST (HIFT) OF 1991

The preceding fifteen years had seen a steady progression in ranges, from Worcester's 25-km reciprocal transmission to the 100-km mesoscale-resolving experiments to Spiesberger's 4000-km climate-oriented transmissions. Forbes and I started to think about an antipodal transmission of 20,000 km[21,22] (the next step would be more difficult).

Birdsall had made some rough estimates that varied by as much as 60 dB; they ranked global transmissions from trivially easy to impossible. The frequency of choice was 57 Hz to avoid electric machinery noise at 50 and 60 Hz. Absorption is not the problem at these low frequencies, varying from 3 dB per 20,000 km in Pacific waters to 5 dB/Mm in Atlantic waters (because of the difference in dissolved salts). Scattering might be. But global acoustics raises new problems. Geometric spreading does not apply on a sphere and in fact great circles *converge* after 10,000 km. Further, Longuet-Higgins had shown that even great-circle ray geometry fails catastrophically within 50 km of the antipole where the departure from a spherical Earth is associated with spheroidal caustics and multiple paths.

For guidance we studied what had been an almost casual addition to a geophysical survey conducted thirty years earlier. While Gordon Hamilton was tending his Bermuda hydrophones, the R/V *Vema* and HMAS *Diamantina* used pressure detonators to fire 300-lb anatol charges near the sound axis off Perth, Australia. These were clearly recorded in Bermuda. In fact, each of the explosions gave a double signature presumably related to the spheroidal caustic (I had previously incorrectly attributed the double signature to a reflection from the Bermuda coastline).

The Oceanographer of the U.S. Navy, Adm. Richard Pittenger, assigned the acoustic source ship *Cory Chouest* to the HIFT. *Cory* had HLF-4 sources suspended through a center well. The sources were limited to an operational depth of 300 m, which dictated their deployment at high latitudes where the sound channel is shallow. Heard Island in the Indian Ocean at 54°S, 74°E filled this requirement, and has the unique location for providing great-circle windows into the North and South Atlantic, North and South Pacific, the Indian Ocean, and Antarctica (Figure 8.6a). (As we later learned, it also has the unique location of a maximum in global wave height statistics.)

Heard Island was discovered by the American Sea Captain Heard in 1853. It is under Australian flag and uninhabited except for occasional visits to study the rich sea life. We chose 0000 GMT 26 January (Australia Day) for a start time, and I personally persuaded colleagues from nine countries (no organization) to be out at sea with receivers suspended into the sound channel.

Birdsall had drawn up the transmission plan: starting on Australia day, 1 hour out of every 3 hours for 10 days; intensity 210 to 220 dB re 1 μPa at 175 m depth:

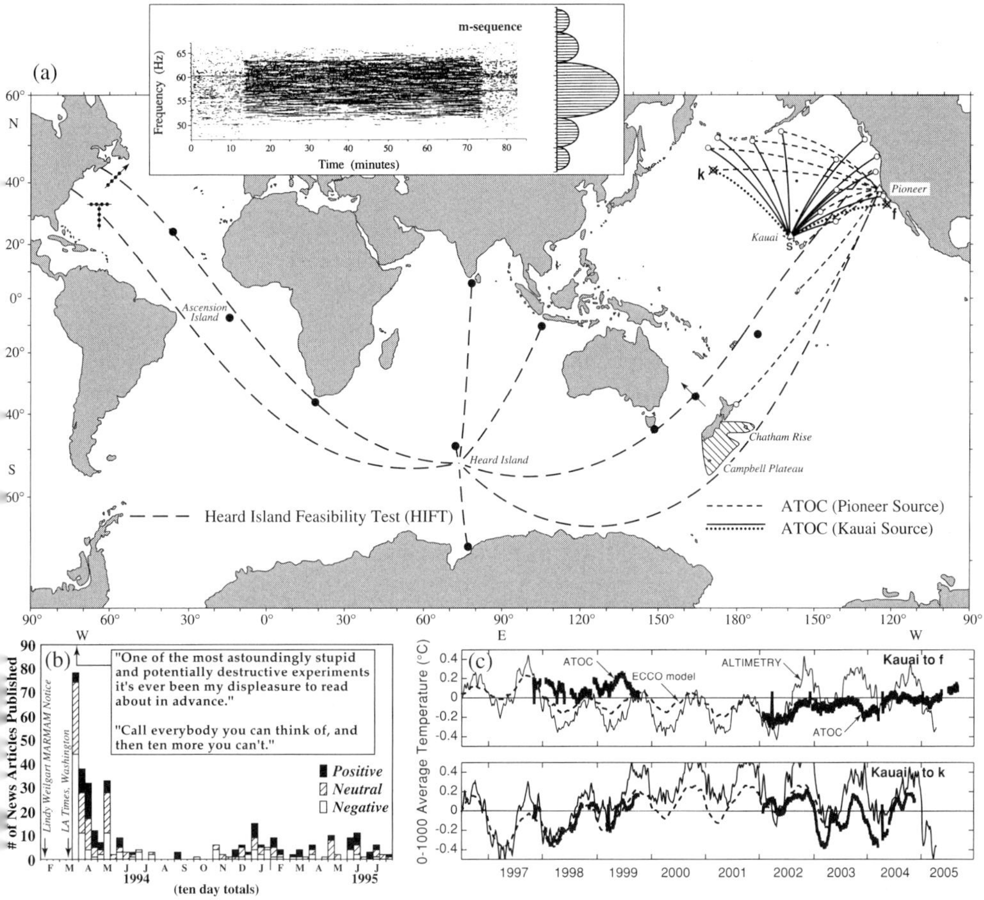

Figure 8.6. (a): Heard Island Feasibility Test (HIFT)[21,22,23] of January 1991. Sources were suspended from the center well of the R/V *Cory Chouest* 50 km south of Heard Island in the Indian Ocean. Black circles indicate receiver sites with single hydrophones, black lines with dots indicate horizontal and vertical receiver arrays. Long dash lines are refracted geodesics (which would be great circles for a homogeneous ocean on a spherical Earth). Signals were received at all stations except for the Japanese station off Samoa and the vertical array at Bermuda (which sank). Inset is signal recorded at Ascension Island; note the "afterglow" of the 57 Hz center frequency following the termination of the 1 hour transmission. Subsequent ATOC transmissions in the Pacific from Pioneer Seamount (short dash) and Kauai (solid and dotted) to various SOSUS arrays are shown. (b): Public response to the proposed ATOC project[25]. The quote in the box is one of the more favorable notices to appear on the MARMAM website dedicated to marine mammals. (c): The mean temperature in the upper 1 km of the northeast Pacific as inferred from satellite altimetry (thin), acoustic thermometry (heavy) and the ECCO model (dashed). From Dushaw (personal communication).

transmitting for 3600 s at 210 dB (2500 W) of acoustic *power* gives about 10^7 J of *energy* per transmission. This is comparable to the 3 x 10^7 J of acoustic energy radiated by the 300-lb explosive charges in the Perth-to-Bermuda transmissions.[22] The coded transmission has the advantage that it spreads this energy over 60 minutes. The procedure for detection is to correlate the received signal with the stored transmission code, leading to a processing gain equal to the degrees of freedom:

$$\text{dof} = \text{bandwidth} \times \text{coherence time} = 10\text{ Hz} \times 300\text{ s} = 3000\ (35\text{ dB})$$

Thus, detection can be achieved at a level well below the ambient sound level and thus inaudible to marine life.

In the midst of preparations we were informed by NOAA's National Marine Fisheries Service that a permit would be required (there had been no such requirement in the preceding 15 years). To make matters worse, the Australian authorities, who had previously supported the test plan without raising the issue of a permit, now decided one would be required. On very short notice, the sister ship *Amy Chouest* with Ann Bowles as chief scientist was tasked to assist the *Cory Chouest* with a parallel program of biological observations. A protocol provided for the termination of the transmissions under certain prescribed circumstances. The biological add-on certainly contributed to the complexity and expenses, yet it provided a welcome partnership with a devoted group of observers who worked under very trying conditions.

Spindel had previously gone to sea on the *Cory* to tune the sources to the lowest attainable frequencies centered on 57 Hz. The two vessels sailed from Freemantle on 9 January with no permits in hand; postponement would have been tantamount to cancellation. Andrew Forbes (an Australian oceanographer) and I were co-chief scientists, Matthew Dzieciuch (a Birdsall Ph.D.) took charge of the acoustic transmissions. We received the American permit while underway one week before the start date, and the Australian permit 21 hours prior to start time. The latter came with a message from the Environmental Minister: "good luck and calm seas." This was not to be.

Our Bermuda listening station and the station north of Seattle were both almost halfway around the Earth, at 18,000 km or $3\frac{1}{2}$ hours acoustic travel time. The reader will recall the 60 dB of *a priori* uncertainty, with little expectation that the signal would be detected at the farthest sites. The evening prior to the scheduled start, technicians aboard the *Cory* requested a five-minute checkout of the sources. Three and one-half hours later we received an puzzled message from Metzger in Bermuda, demanding to know what was going on. Fifteen minutes later a similar message arrived from Birdsall at his station north of Seattle. The signal had arrived on the two opposite American coasts halfway around the Earth, one traveling eastward past New Zealand, the other westward past the Cape of Good Hope. The question of the feasibility of global-scale transmission had been answered, and it was not yet Australia Day. (This was the most exciting moment of my career, and the reader must excuse the detailed reporting.)

The scheduled transmissions commenced on time, and other stations began to report receptions. On 31 January we encountered a gale with 10m seas. One suspended source was torn loose and went to the bottom, and the others were severely damaged. Fortunately there had been 28 successful runs before this untimely termination.

W. Kuperman, the editor of the *Journal of the Acoustical Society of America*, invited us to collect our experience in a special issue,[23] and eighteen papers were published under the heading "The Heard Island papers: a contribution to global acoustics." There were some surprising results. In all the transmissions, the source ship *Cory* had to be underway into winds and waves to avoid being broached. The measured Doppler permits a very accurate determination of the geodetic launch angles (relative to the ship's course). These agreed closely with the *a priori* calculations of the geodetics, except for the path through the Tasman Sea to the American west coast. Evidently the direct route was blocked by unknown bathymetry, but a weak signal got through along an alternate route passing east of New Zealand. For any fixed station the differential Doppler between the early steep and late flat launch angles provided useful tools for ray identification. The conclusion here is that the motion of the source ship (which we regarded as a necessary evil) turned out to be a substantial asset.

Among the oceanographic results is evidence for stripping of high acoustic modes by bathymetric features protruding into the sound channel, with subsequent repopulation. There is strong mode coupling at sharp oceanic fronts (principally the Antarctic Circumpolar front). The final paper by Bowles *et al.* deals with the biological component of HIFT. Observations were restricted to times when the winds were weaker than force 8. The study was severely limited by the short time available for the baseline study prior to transmissions. The evidence suggests some behavioral changes of the marine mammal population, but there was no indication of distress.

HIFT demonstrated that acoustic sources could be detected at antipodal ranges, but the arrival patterns turned out to be too garbled to permit the precise determination of identified travel times. From a climate point of view, the very large ranges are counterproductive since they average across distinct climatic provinces. Following HIFT we have emphasized shorter ranges, generally less than 5000 km.

ACOUSTIC THERMOMETRY OF OCEAN CLIMATE (ATOC), 1993–2000

The completion of HIFT found us in an upbeat mood. We were ready to take on the problem of measuring the change in ocean heat content (central to the ongoing debate on climate change). Our first goal was to monitor the subtropical gyre of the northeast Pacific.[24]

Plans were for a greatly reduced source level (195 dB versus 220 dB at HIFT), at 75 Hz versus 57 Hz center frequency and at midlatitude axial depth of 1000 m versus 175 m HIFT depth. The combined change in source intensity and depth leads to a 35-dB reduction in the sound level within the biologically crucial surface layers. Accordingly we thought that no further permit difficulties would be encountered. This too was not to be.

The Indian Ocean storm was nothing compared to the storm of protest we encountered upon our return. HIFT had been widely publicized under the unfortunate slogan "The shot heard around the World." It caught the attention of the public, but also the attention of the environmental community (Figure 8.6b). On 22 March 1994 an article on the first page of the *Los Angeles Times* suggested that the ATOC transmissions could lead to the death of 250,000 California gray whales. In part this was based on a misunderstanding (195 dB corresponds to 250 watts of acoustic power in water and 250 million watts in air; two months later a correction was published, again on the first page).

Spindel headed the effort to install two bottom-mounted sources; one on Pioneer Seamount off the coast of California, the other off the north coast of Kauai, Hawaii. The Pioneer source was authorized in July 1995 and started transmissions in December. The Kauai source was authorized in February 1996 and started transmissions in October 1997. When the approvals were finally issued, they placed control of the sources under the ATOC Marine Mammal Research Program (MMRP). Climate-oriented research was to be subordinate to studies relating to the effects of sound on marine life. These included a broad array of issues: aerial surveys of marine mammal distribution, elephant seal tagging, playback studies to humpback whales, undersea acoustic recordings to determine changes in humpback vocalization, and so on. None of the studies found any overt or obvious changes. Statistical analysis revealed small changes in the behavior of humpback whales in response to the playback of ATOC-like sounds. The conclusion was that these minor changes would not adversely impact the survival of an individual whale or the North Pacific humpback population.

The end of the period found us with the required permits, but exhausted financially and emotionally. We had attended dozens of hostile public meetings in the unaccustomed role of fighting legal and public relation issues. The Scripps Institution had hired private guards in response to incoming threats. I had the opportunity of briefing Vice President Gore who on a number of subsequent occasions would introduce me (jokingly I hope) as "here comes the whale killer." John Potter[25] has written about "ATOC: Sound Policy or Enviro-vandalism? Aspects of a modern media-fueled policy issue." Jim Lynch of Woods Hole has written a detailed account (unfortunately unpublished) of this difficult period. But there was a good side: we had been exposed to fascinating new problems and forcefully reminded of the unity of the oceanographic sciences.

Still there was significant progress in the physical sciences. Acoustic tomography, though a specialized technique, had touched many branches of physical oceanography. Reciprocal transmissions made it possible to separate barotropic from baroclinic tidal currents with unequaled accuracy. In 1995 Dushaw *et al.* discovered a northward traveling baroclinic M_2 tide that was phase-locked to the surface tides over the Hawaiian Ridge 2000 km to the south.[26] This was subsequently confirmed by satellite altimetry and has been the starting point of the ongoing Hawaii Ocean Mixing Experiment (HOME).

Acoustic tomography is at its best in an integrating mode; integrating horizontally and integrating vertically. As such it does a good job in estimating the variation in total heat content of a basin. The intensity of mesoscale variability greatly exceeds that of climate change, but the acoustic integration over 50 mesoscales reduces mesoscale noise to acceptable levels.

In 1994 the THETIS-2 experiment[27] successfully estimated the seasonal heat content of the western Mediterranean, combining for the first time satellite altimetry with acoustic tomography. In my view the most successful application is the 1994 transmission conducted by Mikhalevsky[28] from a source north of Svalbard, Norway, across the pole to a receiver near Pt. Barrow, Alaska. An ultralow-frequency source (20 Hz) was dictated by the scattering loss from the ice cover. An earlier than expected arrival of acoustic mode 2 (whose mode function peaks at the depths occupied by the North Atlantic Intermediate Water) implied a warming by 0.4°C. A subsequent 1999 transmission implied further warming by 0.5°C. The results are roughly consistent with independent CTD surveys conducted from submarines. The polar environment provides three advantages to acoustic tomography (aside from being available in winter when it is not accessible to standard oceanographic techniques): a low background of internal waves (the ultimate limit to precision), a dispersive sound channel favorable to vertical resolution, and remoteness from public attention.

Parallel efforts by French, German, and Japanese oceanographers during this period were less restricted by the biological consideration. An international working group (SCOR 96) under the chairmanship of David Farmer met in France, India, Denmark, Austria, and Japan.[29]

NORTH PACIFIC ACOUSTIC LABORATORY (NPAL), 2000–

A summary of this period has been prepared by Worcester *et al.*[30] The ATOC Kauai transmissions had lasted from October 1997 to October 1999. Worcester then began the arduous process of authorization and the Kauai source finally resumed operation in January 2002 as part of the NPAL program. The Kauai time series is therefore now seven years long, albeit with a substantial gap.

From the early days we had practiced a joint analysis of satellite altimetry and acoustic tomography. This combined the good horizontal resolution of altimetry with the depth resolution and good time resolution of tomography. What was missing was a good model framework for the combined analysis. The ECCO Consortium (Estimating the Circulation and Climate of the Ocean) finally provided the vertical resolution needed for calculating the acoustic propagation through the model ocean (the forward problem), allowing straightforward comparison of measured and predicted travel times. The model assimilates a variety of satellite and *in situ* data, including TOPEX-POSEIDON altimetry, WOCE hydrography, XBT sections, and Argo floats.

The next step will be to use the acoustic travel times as integral constraints on the model variability. Using modern ocean state estimation methods, the acoustic data can be compared to, and ultimately combined with, the upper ocean data to detect changes in abyssal ocean temperature, and to detect whether the various data types are complementary. A typical Argo volume-mean is subject to a 0.15°C uncertainty. The corresponding uncertainty in the ATOC-derived mean is typically only 0.02°C, a direct result of the suppression of mesoscale variability by the acoustic path averaging.

The seven-year time series (Dushaw, personal communication) reveals a surprising degree of inhomogeneity within the northeast Pacific (Figure 8.6c). From Kauai northeastward toward California (receiver f) there is a modest cooling trend from 1997 until the present time; the path to the northwest (receiver k) shows modest warming from 1997 to 2003, followed by an abrupt cooling. The changes are related to a warming of the central Pacific during this interval. The climate variability on a 10 Mm-decadal space-time scale overwhelms a global secular trend. Perhaps this should have been foreseen; the disappointing message is that our records are too local and far too brief to contribute to the ongoing debate about "global" warming.

THE FUTURE

In retrospect, the program has suffered from permitting problems, a high start-up cost, and the mismatch in the time scales of climate and funding cycles. There is some hope for a partial relief.

At the International Conference on the Ocean Observing System for Climate: OCEANOBS 99 in St. Raphael, France (18–22 October 1999), Dushaw *et al.*[31] summarized the appropriate role for acoustic tomography in observing ocean climate. Some preliminary planning had previously been done for thermometry systems in the Atlantic, Arctic, and Indian oceans. Plans are underway for a global ocean observing system. As part of such a system, ORION is to provide a coarse network of about two dozen deep-sea moorings, the successors to the weather ships of the early twentieth century (Figure 8.7). The moorings would be instrumented with a diverse array of meteorological and geophysical sensors. We have proposed that acoustic transceivers at the ORION moorings should monitor the oceans *between* the moorings. This would lead to the long-term global exposure required for monitoring the response of the oceans to greenhouse warming and other long-period forcing.

We envision vertical transceiver arrays plugged in at the bottom of the ORION moorings. The ORION system would provide for electric power, a millisecond time base, data transmission via satellite link, plus occasional servicing. This would meet a large fraction of the present high start-up cost of acoustic tomography deployments.

The primary task is to provide data on ocean warming; preliminary estimates indicate that deep ocean warming by a few millidegrees could be detected. There is at present no information on possible large-scale long-term trends between 2000 and 5000 m depths.

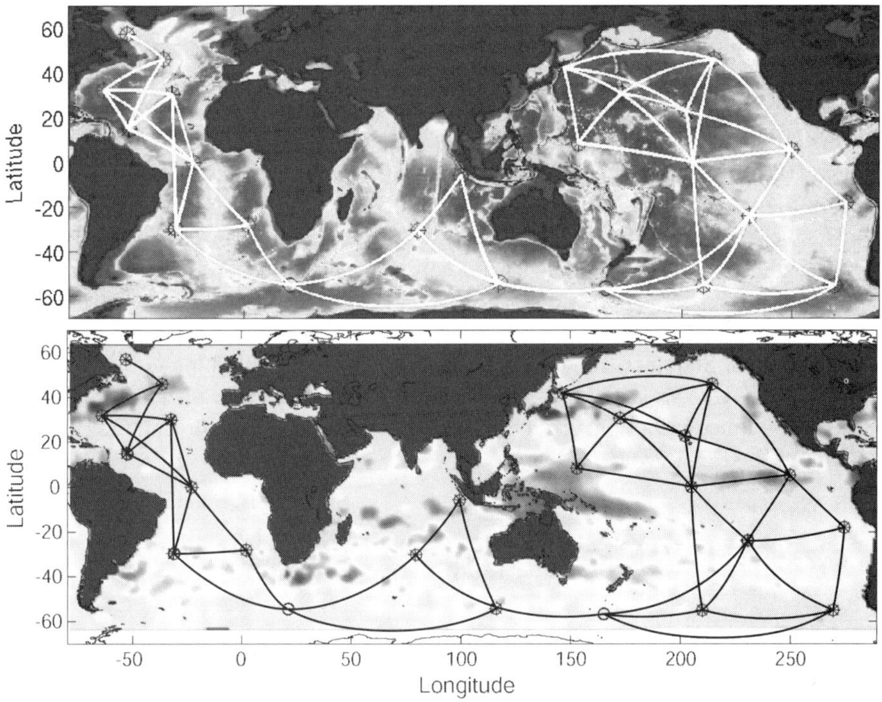

Figure 8.7. A straw-man global mooring array. The lines are proposed acoustic paths between mooring transceivers. The top figure is plotted against the background of the major bathymetric features. The bottom figure is plotted against a background of estimated temperature variability. From Dushaw and Worcester (personal communication).

But much more could be done. Sound rays that are reflected at the sea surface or turn in the upper ocean layers (the SOFAR "overture") are sensitive indicators of mixed layer formation and other upper ocean processes. This information can be read at the lower turning points, typically at 3 to 4 km depths. As this story is being written, we are studying the possibility of abyssal arrays plugged in to ORION moorings at the sea bottom and extending to the axis (but no further), thus taking advantage of the relative calm and benign environment at the *conjugate* depths.

To carry out these plans will require the support of the community of physical oceanographers. Here the record is mixed. Traditional members of this community are not excited about the oceanography but find the application of inverse methods interesting. Modelers consider our application of inverse theory as routine but the oceanography interesting. I hesitate to comment about the future; predictions are not likely to be followed if previous experience is any guide. But it is inconceivable that oceanographers (like other marine mammals) should not take advantage of the fact that the ocean is transparent to sound.

ACKNOWLEDGMENTS

I have lifted freely from papers by Peter Worcester, Robert Spindel, and Carl Wunsch; we have been partners since the earliest days of ocean acoustic tomography, and without their participation none of this story would have been written. The authors (Munk, Worcester, and Wunsch) of the 1995 book *Ocean Acoustic Tomography* dedicated their volume to D. Behringer, T. Birdsall, M. Brown, B. Cornuelle, R. Heinmiller, R. Knox, K. Metzger, J. Spiesberger, R. Spindel, and D. Webb. Additional persons cited in a 2005 review of "Acoustic remote sensing of ocean gyres" by Worcester, Munk, and Spindel are A.B. Baggeroer, C. Clark, J. A. Colosi, D. Costa, B. D. Dushaw, M. A. Dzieciuch, A. M. G. Forbes, C. Hill, B. M. Howe, J. Marshall, D. Menmenlis, J. A. Mercer, and D. Stammer. David Farmer chaired the SCOR Working Group 96 from 1994 to 1998. Our research has been supported over the years by the Office of Naval Research, with contributions from the National Science Foundation and the Defense Advanced Research Project Agency.

FOOTNOTES AND REFERENCES

We shall refer to previous summaries of the subject for references and further information.

MWW pxyz refers to page xyz of Munk, W., P. Worcester, and C. Wunsch, 1995. *Ocean Acoustic Tomography*. Cambridge University Press, 450 pp.

MWS refers to Munk, W. H., P. F. Worcester, and R. C. Spindel, 2003. Acoustic Thermometry of Ocean Climate. Preface to *Collected Papers, Acoustic Thermometry of Ocean Climate (ATOC)*.

WMS refers to Worcester, P. F., W. H. Munk, and R. C. Spindel, 2005. Acoustic remote sensing of ocean gyres. *Acoustics Today*, in press.

1 Munk, W., and C. Wunsch, 1982. Observing the ocean in the 1990's. *Philos. Trans. R. Soc. London* **307**, 439–464.

2 Krümmel, O., 1911. *Handbuch der Ozeanographie*. J. Englehorn, Stuttgart: Vol. 1 526 pp, Vol. 2 764 pp.

3 Sverdrup, H. U., M. W. Johnson, and R. H. Fleming, 1942. *The Oceans: Their Physics, Chemistry and General Biology*. Prentice–Hall, Englewood Cliffs, NJ, 1087 pp.

4 Defant, A., 1961. *Physical Oceanography*. Pergamon Press, New York. Vol. 1 729 pp, Vol. 2 598 pp. First completed in German in 1945.

5 MWW p356–8.

6 Hamilton, G. R., 1977. Time variations of sound speed over long paths in the ocean. In: *International Workshop on Low-Frequency Propagation and Noise* (Woods Hole, MA, 14–19 October 1974), 7–30. Department of the Navy, Washington, DC. See MWW p331.

7 MWW p358–9.

8 Garrett, C., and W. Munk, 1972. Space-time scales of internal waves. *Geophys. Fluid Dyn.* **2**, 225–264. Munk, W. H., and F. Zachariasen, 1976. Sound propagation through a fluctuating stratified ocean: Theory and observation. *J. Acoust. Soc. Am.* **59**(4), 818–838.

9 Flatté, S. M., R. Dashen, W. Munk, K. Watson, and F. Zachariasen, 1979. *Sound Transmission Through a Fluctuating Ocean*. Cambridge University Press, London.

10 Worcester, P. F., 1977. Reciprocal acoustic transmission in a midocean environment: Fluctuations. *J. Acoust. Soc. Am.* **62**, 895-905. See MWW p360.

11 Spiesberger, J. L., R. C. Spindel, and K. Metzger, 1980. Stability and identification of ocean acoustic multipaths. *J. Acoust. Soc. Am.* **67**, 2011–2017. See MWW p362.

12 Munk, W., and C. Wunsch, 1979. Ocean acoustic tomography: A scheme for large scale monitoring. *Deep-Sea Res.* **26A**, 123–161.

13 Munk, W., and C. Wunsch, 1982. Up/down resolution in ocean acoustic tomography. *Deep-Sea Res.* **29**, 1415–1436. Munk, W., and C. Wunsch, 1983. Ocean acoustic tomography: Rays and modes. *Rev. Geophys. Space Phys.* **21**(4), 777–793. Munk, W., and C. Wunsch, 1985. Biases and caustics in long-range acoustic tomography. *Deep-Sea Res.* **32**, 1317–1346. Munk, W., and C. Wunsch, 1987. Bias in acoustic travel time through an ocean with adiabatic range-dependence. *Geophys. Astrophys. Fluid Dyn.* **39**,1–24.

14 Munk, W., and C. Wunsch, 1982. Observing the ocean in the 1990's. *Philos. Trans. R. Soc. London* **307**, 439–464.

15 See footnote MWW p363.

16 See MWW p363–381.

17 Ocean Tomography Group, 1982. A demonstration of ocean acoustic tomography. *Nature* **299**, 121–125. See MWW p299. (Members of the Ocean Tomography Group are D. Behringer, T. Birdsall, M. Brown, B. Cornuelle, R. Heinmiller, R. Knox, K. Metzger, W. Munk, J. Spiesberger, R. Spindel, D. Webb, P. Worcester, and C. Wunsch.)

18 Spiesberger, J. L., T. G. Birdsall, K. Metzger, Jr., R. A. Knox, C. W. Spofford, and R. C. Spindel, 1983. Measurements of Gulf Stream meandering and evidence of seasonal thermocline development using long range acoustic transmissions. *J. Phys. Oceanogr.* **13**, 1836–1846. See MWW p365.

19 Spiesberger, J. L., and K. Metzger, 1991. Basin–scale ocean tomography: A new tool for studying weather and climate. *J. Geophys. Res.* **96**, 4869–4889. Spiesberger, J. L., and K. Metzger, 1992. Basin–scale ocean monitoring with acoustic thermometers. *Oceanography* **5**, 92–98.

20 Worcester, P. F., J. F. Lynch, W. M. L. Morawitz, R. Pawlowicz, P. J. Sutton, B. D. Cornuelle, O. M. Johannessen, W. H. Munk, W. B. Owens, R. Shuchman, and R. C. Spindel, 1993. Evolution of the large-scale temperature field in the Greenland Sea during 1988–89 from tomographic measurements. *Geophys. Res. Lett.* **20**, 2211–2214. Sutton, P. J., W. M. L. Morawitz, B. D. Cornuelle, G. Masters, and P. F. Worcester, 1994. Incorporation of acoustic normal mode data into tomographic inversions in the Greenland Sea. *J. Geophys. Res.* **99**, 12487–12502. See MWW p347-8. Pawlowicz, R., J. F. Lynch, W. B. Owens, P. F. Worcester, W. M. L. Morawitz, and P. J. Sutton, 1995. Thermal evolution of the Greenland Sea Gyre in 1988–89. *J. Geophys. Res.* **100**, 4727–4750. Morawitz, W. M. L., P. J. Sutton, P. F. Worcester, B. D. Cornuelle, J. F. Lynch, and R. Pawlowicz, 1996. Three-dimensional observations of a deep convective chimney in the Greenland Sea during winter 1988/1989. *J. Phys. Oceanogr.* **26**, 2316–2343. Sutton, P. J., W. M. L. Morawitz, P. F. Worcester, and B. D. Cornuelle, 1997. Temperature evolution of the upper ocean in the Greenland Sea January to March 1989. *J. Geophys. Res.* **102**, 27, 861–27,874.

21 Munk, W. H., and A. M. G. Forbes, 1989. Global ocean warming: An acoustic measure? *J. Phys. Oceanogr.* **19**, 1765–1778.

22 MWW p337–9, p344–5.

23 Vol. 59, No. 4, October 1994 of the *J. Acoust. Soc. Am.*

24 62 ATOC research papers 1992–2003 and 15 arctic tomography papers have been assembled by P. Worcester in four volumes. The first volume has a brief chronology and a preface by W. Munk, P. Worcester, and R. Spindel (MWS).

25 Potter, J. R., 1994. ATOC: sound policy or enviro-vandalism? Aspects of a modern media-fueled policy issue. *J. Environ. Dev.* **3**, 47–62. See MWW p379.

26 Dushaw, B. D., B. D. Cornuelle, P. F. Worcester, B. M. Howe, and D. S. Luther, 1995. Barotropic and baroclinic tides in the central North Pacific Ocean determined from long-range reciprocal acoustic transmissions. *J. Phys. Oceanogr.* **25**, 631–647.

27 THETIS Group, 1994. Open–ocean deep convection explored in the Mediterranean. *EOS, Trans. Am. Geophys. Union* **75**, 219–221. (Members of the THETIS Group are F. Schott, U. Send, G. Krahmann,

C. Mertens, M. Rhein, M. Visbeck, Y. Desaubies, F. Gaillard, T. Terre, J. Papadakis, M. Taroudakis, G. Athanassoulis, and E. Skarsoulis.)

[28] Mikhalevsky, P. N., A. N. Gavrilov, and A. B. Baggeroer, 1999. The Transarctic Acoustic Propagation experiment and climate monitoring in the Arctic. *IEEE J. Oceanic Eng.* **24**(2), 183–201.

[29] We refer the reader to the Proceedings of the International Symposium on Acoustic Tomography and Acoustic Thermometry held in Tokyo in February 1999 under the auspices of the Japan Marine Science and Technology Center (JAMSTEC).

[30] Worcester, P. F., W. H. Munk, and R. C. Spindel, 2005. Acoustic remote sensing of ocean gyres. *Acoustics Today,* in press.

[31] Dushaw, B. D., G. Bold, C. -S. Chui, J. A. Colosi, B. D. Cornuelle, Y. Desaubies, M. A. Dzieciuch, A. M. G. Forbes, F. Gaillard, A. Gavrilov, J. Gould, B. M. Howe, M. Lawrence, J. F. Lynch, D. Menemenlis, J. A. Mercer, P. Mikhaelvsky, W. H. Munk, I. Nakano, F. Schott, U. Send, R. C. Spindel, T. Terre, P. F. Worcester, and C. Wunsch, 2001. Observing the ocean in the 2000's: A strategy for the role of acoustic tomography in ocean climate observation. In: *Observing the Oceans in the 21st Century*, C. J. Koblinsky and N. R. Smith, eds., GODAE Project Office, Bureau of Meteorology, Melbourne, 391–418.

Chapter 9

A History of Thermocline Theory

JOSEPH PEDLOSKY

INTRODUCTION

The focus of my attention in this essay is to give the reader some insight into the development of the theoretical ideas concerning the thermocline from the very personal point of view of someone involved in that development. It is not a review paper of the full scientific history of the subject or a detailed scientific discussion of the problem. Rather, I want to present the more personal side of the history of the attempt to reach an understanding of the phenomenon of the thermocline. In that special sense this is a personal memoir of my involvement in that quest and I hope to capture the flavor of that experience for others. It culminated for me in the development of the theory of the Ventilated Thermocline, a theory that evolved in collaboration with Henry Stommel and Jim Luyten and so the story is one person's view of what was a collegial effort.

The presence of the thermocline, the region of rapid increase of density with depth in the first one or two kilometers of the subtropical oceans, has been recognized for a long time. Yet looking back at the older literature it is hard to find a reference to the *problem* of the thermocline, that is, in explaining the maintenance of this sharp density gradient or its cause. The thermocline itself is bowl-shaped in the meridional plane and contains fairly entirely the wind-driven currents of the major subtropical gyres

Joseph Pedlosky, Senior Scientist, Woods Hole Oceanographic Institution.

so there is much to explain. It is provocative to note that the very word does not even appear in the index of that great tome, *The Oceans*, by Sverdrup, Johnson, and Fleming (1942) which was for a long time the definitive work on oceanography although the word appears a few times[1] in the text. More frequently, the thermocline is referred to there as a "discontinuity layer" or "transition layer." This makes the thermocline appear as a curious secondary feature separating zones of greater oceanographic interest. Similarly in Defant's two-volume treatise on physical oceanography (1961), the thermocline is noted largely as a transition zone between what are supposed to be oceanographic analogues of the atmosphere's troposphere and stratosphere. The existence of such a transition layer is taken, misleadingly, as a natural process similar to the seasonal thermocline in lakes. Indeed, the first use of the word "thermocline" appears in the limnology literature of the late nineteenth century, a fact for which I am indebted to my colleague Bruce Warren.

It does seem like one of those examples, not uncommon in science, in which a phenomenon is noted but that it might be a *problem* requires first that it be posed as a problem, i.e., why should there be a permanent thermocline in the ocean? It is a bit like becoming used to looking at the Rocky Mountains and taking their presence for granted. It may not be natural for most people in their neighborhood to ask why they are there but once the question *is* asked it is easy to realize the importance of finding the answer.

Of course it is a danger of amateur historiography to overlook past insights into a scientific problem that contributed to a background understanding but which often enter into oblivion because these insights were not united to a powerful enough method to obtain a complete solution to the problem. So, for example, the insightful work of Montgomery (1938), in which it is pointed out that the water mass properties at depth in the subtropical gyre can be traced back along an isopycnal to the properties of surface water, was an idea clearly influenced by the meteorological interest of the time in isentropic analysis of atmospheric motions suggested by Rossby (1937). This idea is seen in even more explicit form in the interpretive work of Iselin (1936, 1939) whose much reproduced schematic shows explicitly how water from the mixed layer sliding down along isopycnal surfaces sets the vertical distribution of density (and so implicitly also its horizontal distribution) in the subtropical gyres. These very profound insights were, however, innocent of a powerfully enough unifying dynamical foundation that could go beyond a quasi-diagnostic explanation to provide the basis for a theoretical understanding of thermocline structure. Interestingly enough, the key ingredient has turned out to be potential vorticity conservation and this was much on the minds of Rossby and Montgomery who had collaborated on its application to atmospheric motions. A similar application to oceanic motion had to wait another 40 years.[2]

---

[1] My colleague, Bruce Warren, has counted ten appearances of the word in the book.

[2] I am excluding the remarkable work of Welander, discussed below.

In the years that followed there were few attempts to deal with the structure of the thermocline except for some linear models (Stommel and Veronis, 1957; Pedlosky, 1969; Gill, 1985) that attempted to describe the capture of the wind-driven motion to an upper region of the ocean while accepting an already specified background stratification as a starting point for the linearization of the theory. Such treatments, interesting in their own right, are clearly incapable of addressing the fundamental question of the density structure that defines the thermocline. They do, however, have the virtue of indicating the beta-effect as being the important dynamical mechanism for limiting the depth of penetration below the upper sea surface of the wind-driven motion.

The principal issue is easily stated. The ocean is heated nonuniformly but persistently at the sea surface. How is it, then, that the signature of that forcing penetrates to only about 20% of the total ocean depth even though the heating has gone on for millions of years? Clearly, a dynamical process is required to trap the thermal signal to the upper ocean, but what specifically is the operating mechanism?

MODERN THEORIES, 1959 TO 1980

It is a weakness of human perception to define "modern," in the sense of "advanced," as that which is simply contemporaneous with one's own time. Nevertheless, I think it is fair to mark the modern era of thermocline theory with the advent of two remarkable papers that appeared simultaneously (Robinson and Stommel, 1959, and Welander, 1959) that did denote a real step forward in development. As remarkable as the papers, published back-to-back in an issue of *Tellus*, are themselves for their pioneering approach to the problem of the thermocline, it is the foreword to the first paper (and serving as a foreword to both) that immediately captures the reader's attention. The foreword alerts the reader to the inconsistency of the basic physical assumptions of the two papers, declines to suggest a test that would distinguish which is correct (since they both seemed to predict a reasonable thermocline structure), and leaves it to the reader to choose between them. The Robinson and Stommel paper focused on an explanation of the thermocline as a lake-like balance between the downward diffusion of heat from the surface and the advection of cold water. It is interesting to note that in the model the zonal variation of temperature was ignored along with the accompanying zonal advection of temperature. As we know, a surface condition in which the temperature is independent of longitude does not, in fact, produce a solution independent of longitude but that assumption allowed the authors to find a similarity transformation, which, after a type of linearization, led to solutions for the density anomaly which exponentially decreased with depth. One of the most interesting aspects of the paper is its emphasis on obtaining an estimate of the vertical velocity at the base of the thermocline in the hope of establishing a rational basis for the driving mechanism of the abyssal circulation as proposed by Stommel (1958). Another point of interest, in hindsight, is the discussion of the size of the turbulent diffusion coefficient, κ, needed by the model. The authors note that an extremely large value would turn the thermocline into an unrealistic, linear top-to-bottom profile of temperature while a very small value of κ, less than 1 cm^2/sec, would lead to an advective model of the thermocline which, using the similarity form introduced by the authors, is claimed to provide the wrong form of the latitude dependence of the thermocline. This I believe is an unfortunate consequence of ignoring the zonal advective term.

Welander's approach, on the other hand, strikes the contemporary reader as being remarkably prescient. He points out that "the importance of diffusion processes in large-scale ocean dynamics has not yet been proved" and he suggests ignoring diffusion altogether. Doing so, he obtains a purely advective model described by a nonlinear partial differential equation for a variable related to the vertical integral of the pressure anomaly, itself related to an integral of the density, his M function. The difficult, resulting equation has a simple solution in which the density field exponentially approaches its uniform abyssal value with a scale height that is directly proportional to the Coriolis parameter. We recognize today that this solution is one in which the planetary potential vorticity (the Coriolis parameter multiplied by the vertical temperature gradient) is constant on density surfaces so the adiabatic advection of density and potential vorticity are automatically satisfied. Welander extended his advective model in a series of later papers (Welander, 1971a,b) to include more complex dependence of the potential vorticity on Bernoulli function[3] as well as density. My own personal view is that Welander had a deeper insight, shown especially in the second of these two 1971 papers, than almost all his contemporaries. He anticipated a thermocline in which diffusion was important only on the boundary of an adiabatic region separating the ideal-fluid thermocline from the abyss and gave clear scaling arguments and a cogent physical discussion of the way in which the model's parts should fit together. Again, what was lacking was a theoretical tool to provide a basis for determining the functional relationship between potential vorticity, density, and Bernoulli function. Rather than pursuing this line further, Welander and Robinson collaborated on a model in which the advective/diffusive balance was added back to Welander's formulation. This leads to a very complex nonlinear partial differential equation. Only special similarity solutions of this kind of model were ever found, i.e., models whose functional behavior with depth is the same everywhere except insofar as the vertical coordinate is stretched as a function, in the general case, of latitude and longitude. Veronis (1969) gives a clear review of this theoretical approach and its results. Many were the authors who approached the problem in this form in an attempt to obtain a physical understanding of the mechanics that shape the thermocline. Perhaps the most distinguished of these is the formulation and solution of Needler (1967) but his solutions, again of similarity type, were mostly of the constant potential vorticity form and shared with all the other models of this type the inability to deal with a simple forward problem. Namely, if the surface density (or temperature) field

[3] The planetary scale Bernoulli function $p + \rho g z$ is also called the Montgomery stream function.

is given as well as the surface Ekman pumping vertical velocity, what is the resulting shape of the density surfaces, what determines the depth of the thermocline, and how is the motion field determined? As a problem in fluid mechanics this is clearly what is desired rather than finding a fortuitous solution of the differential equations and asking for the consistent boundary conditions that allow it.

More to the point, the solutions with constant potential vorticity were simply not able to answer the simple question, "How important is κ for determining the thermocline structure?" In the similarity solutions constant potential vorticity is a solution for *both* the ideal fluid model *and* the model with dissipation and the only difference between them is the vertical velocity predicted at the base of the thermocline and that depends linearly on κ.

Again, in my own opinion, the problem of the thermocline as a problem in fluid mechanics was stymied by the prospect of dealing with a complex mathematical problem that transformed thermocline theory into a rather abstruse mathematical search for special solutions to a very difficult differential equation. And yet, also in my opinion, Welander had in a very thoughtful way outlined a fruitful adiabatic approach to the problem that was not followed up. It is an interesting question in my own mind why this is so. As I recall the years of the 1960s as a student and then a young researcher in the field, the power of boundary layer theory to deal with fundamental and difficult problems in ocean circulation theory had seemed indisputable. At the same time there was an intuitive feeling that the thermocline was a *thermal* boundary layer and it was natural to think of it as an advective/diffusive one. Once diffusion enters the problem on the same level as advection, and the three-dimensional character of the dynamics is respected, it becomes almost impossible to make theoretical progress in a way that leads to deeper physical understanding. One has to admit in retrospect that, as a consequence, progress seems to have stalled for at least a decade and most people, including me, accepted the similarity solutions as the closest to a theoretical basis for understanding the thermocline as we were going to get. It is also provocative and natural to ask why numerical models of the thermocline, not limited by the analytical difficulties described earlier, were not able to confront the physical issues and elucidate the basic physics of the problem. There are probably many reasons for this including a desire to deal with other problems, especially the problem of the oceanic eddy field at that time, or to focus on increasingly complex models that could simulate important oceanographic processes, but the fact remains that conceptual help in understanding the thermocline came first from simple analytical models.

LAYER MODELS OF THE THERMOCLINE: ONE STEP BACK, TWO STEPS FORWARD

In August 1980 Peter Rhines gave a seminar in Woods Hole shortly after his return from a sabbatical visit in Cambridge. The title of his seminar, in which Bill Young was listed as a second author, was "Simple theoretical models of the circulation." There was no hint that Peter and Bill had been working on thermocline theory and, in fact, the seminar started with an elementary, although incisive, review of Sverdrup theory for the midocean. Abruptly, the seminar shifted focus to a rather fundamental question. If one considers the region of the thermocline as consisting of a sequence of frictionless, immiscible layers, i.e., an ideal fluid, what would drive into motion any layer not immediately in contact with the overlying Ekman layer? It would appear that only the uppermost layer would carry the total Sverdrup transport. Then, if one constantly increased the number of such layers, it would appear that the Sverdrup transport would be confined to an increasingly narrow layer near the sea surface as the resolution of the model increased with correspondingly large velocity. This *reductio ad absurdum* emblemizes the fundamental physical problem of thermocline theory, that is, the mechanism and the scale of penetration of the surface wind forcing. If the more obvious question to ask is why the surface forcing should be arrested at a vertical scale short of the full ocean depth, an equally fundamental question, if more subtle, is how the forcing can penetrate at all to a domain in which frictional stresses or thermal turbulent diffusion are not important. It was this question that Rhines and Young addressed and their solution was brilliant and elegant. If each of the layers used to resolve the region of the thermocline becomes thinner and thinner as more layers are added to increase the vertical resolution of the model, the uppermost layer containing all the Sverdrup transport will have large velocities, so large that the isolines of potential vorticity of the layer beneath, defined as the Coriolis parameter divided by the layer thickness, f/h, would be grossly distorted from latitude circles as the upper interface of the layer responds, by the thermal wind balance, to the strong velocities. If the potential vorticity trajectories distort enough so that they avoid the eastern boundary, it is possible that a free mode of geostrophic motion can close on those distorted contours. Even an infinitesimal forcing, say by an eddy flux of potential vorticity, can then resonate with that free mode and drive an order one motion in the layer previously considered at rest. That motion in the second layer would alter the potential vorticity of the layer beneath *it* and so on, and the motion would burrow down to a depth that depended on the wind forcing and the layer thicknesses. Under some rather plausible assumptions of the nature of the potential vorticity flux, Peter and Bill deduced that the subsurface motion in the layers not in direct contact with the Ekman pumping would have uniform potential vorticity. At one stroke, this provided both an explanation of the depth wind-driven motion and a method for its calculation (Rhines and Young, 1982a,b). Moving to a layered model, rather than a continuous one, might seem a step backward but the resulting physical insights produced an enormous advance in our understanding of the physics of the thermocline.

I remember the very strong impression Peter's seminar made on me. Not only was the theory itself extremely elegant but the thought that came most forcefully to my mind was that a sense of physics had been restored to the search for an understanding of the thermocline. I remember that Hank Stommel, sitting beside me in the lecture, shared that view as well. And, as most working scientists will appreciate, a sense of envy was mixed with the admiration because one always regrets missing out on an idea that seems so obvious but only in retrospect. I recall that in leaving the seminar I mentioned to Stommel that the theory seemed to me very powerful, indeed it seemed to explain, and perhaps had been stimulated by, some earlier modeling experiments Peter had done with Bill Holland (1980), but one feature of the oceanic thermocline that it lacked was the outcropping of the isopycnal surfaces within the subtropical gyre. In the Rhines and Young quasi-geostrophic model each layer was perturbed only slightly from its rest thickness and so in the quasi-geostrophic layer model the surface density was a constant. This seemed to me the main direction in which the Rhines–Young theory could be extended further.

At the time I was involved in some other work and did not do anything to pick up on this suggestion but, if my memory serves correctly, nearly a year later Stommel came to my office to show me the results of some numerical calculations he and Jim Luyten were doing for a two-layer model over a resting abyss. The model had a single outcrop and they were numerically finding the structure of the interface heights as a function of latitude; the longitudinal dependence could be independently determined from the Sverdrup relation. Hank had been coming regularly to talk to me about his ideas during the year since I had arrived at Woods Hole from Chicago. The discussions were always deeply interesting; Hank, after all, had a prodigiously creative mind. Usually, I could provide only encouragement or ask for some clarification but no further collaboration followed. In this case, however, my overall interest in the problem and the chance to formulate a simple model of a difficult physical process really appealed to me. After Hank left my office I brushed aside what I had been working on and took up the analytical formulation of the problem. After working out the geostrophic pressure fields in terms of the layer thicknesses of the $2\frac{1}{2}$-layer model, I confronted the essential issue. The lower layer north of the outcrop line is directly exposed to Ekman pumping. In the subtropical gyre it is forced southward according to the Sverdrup relation and when meeting the outcrop line it becomes shielded from the Ekman pumping by the upper, lighter layer. What would happen to the fluid in the lower layer next? Clearly it was being pushed beneath the upper layer. The word "subduction" came immediately to my mind. In Chicago I had worked with Frank Richter, a student in the Geophysical Sciences Department, and we had been thinking hard about mantle convection, downgoing slabs, and a model of that process had formed Frank's Ph.D. thesis. This seemed an oceanographic equivalent except that for the subducted water its potential vorticity would be conserved along its (unknown) streamline path. For this layer the potential vorticity was just f/h_2 where h_2 is the thickness of the second layer and that quantity would have to be constant on a streamline that, because of geostrophy, would be along a line of total depth $h = h_1 + h_2$ so that $f/h_2 = Q(h)$ where Q is some yet to be determined function. This, the determination of Q, was the nub of the problem. As a student I had been tremendously impressed by Charney's inertial model of the Gulf Stream (Charney, 1955) in which the functional relationship between the sum of the planetary and relative vorticity to the external Sverdrup streamfunction needed to be determined.

Generally, the determination of these relationships involves the inversion of a rather complex functional relationship and it is rare that a simple form emerges. However, as I stared at the relation $f/h_2 = Q(h)$ it suddenly occurred to me that at the latitude of the outcrop h_2 was in fact equal to h since h_1, *by definition*, vanishes there. Further, if the outcrop line is a line of constant latitude where f is a constant, say f_2, the potential vorticity of every fluid column in layer 2 along the outcrop line would just be f_2/h and that relation would be preserved for all streamlines issuing from the outcrop line so that the function Q was simply f_2/h.[4] I remember being startled by the utter simplicity of the result. It hardly seemed possible to me that it could be so simple. Assuming it was correct, it allowed each layer thickness to be simply related to the total depth, h, and the Sverdrup relation then gave a direct result for the field of h and so h_1 and h_2 over the gyre. It was a matter of 20 minutes to write up the result in a clean copy that I brought down to Hank's office. As luck would have it, he was not there so I left the clean copy on his desk and went back to my office. I was still in a state of some excitement when Hank called me and thanked me for the notes but said my result was clearly wrong. According to their numerical calculations it appeared to him that the thermocline depth shallowed exponentially as the equator was approached and my solution became shallow only algebraically. Well, I could not see the error in the calculation but it *was* so simple that I confess to having had a residue of doubt that such a simple result would solve a problem that had bedeviled us for so long. So I just decided to go home, sleep on it that night, and check the result again in the morning.

I did not have to wait. Later that same evening, in the first of what would be a series of after-hours telephone conversations, Hank called and said that he had plotted the analytical solution over the numerical solution he and Jim had obtained and they matched perfectly. Could we talk about it in the morning? Well, one can just imagine my state of mind at that moment.

The next day Hank and I spoke about the analytical solution. Jim had grasped it immediately and had explained it to Hank and I sensed that Hank was both excited and a little disappointed that the numerical work was no longer necessary. Indeed, he wondered whether the analytical solution was intrinsically limited to the two-moving-layer model and that is what led to dealing with the three-layer model that later formed the basis of the Ventilated Thermocline paper (Luyten *et al.*, 1983). In retrospect, Hank and I agreed that the development of the three-layer model was probably a poor decision. The basic idea of the model is already captured in the two-layer version and three layers added complexity with a rather small return in understanding. However, with the model in hand we were able to think about what aspects of the boundary conditions and forcing determined the structure.

---

[4] Ironically, the case of zonal distribution of temperature, although the easiest in this case, was not possible for the similarity solutions.

A key feature of this first model was that all layer thicknesses vanished on the eastern boundary to satisfy the no-normal flow condition there. This particularly troubled Hank and he insisted that there ought to be some way to save a finite layer thickness on the eastern boundary. Why couldn't the subducted layer just be motionless near the eastern boundary? It was a good question like all of Hank's were and due to his insistence I was moved to try to find a solution that could accomplish that. I settled on knitting together two solutions, one in which, for a two-layer model, the second layer was first arbitrarily at rest near the eastern wall and one in which the lower layer was in motion due to the subduction from the outcrop line and then sought to satisfy the compatibility condition that the layer thicknesses be continuous at the joining line between them. Of course it turned out that the joining line was just the trajectory of the easternmost path issuing from the outcrop line on the eastern boundary so that the total solution had a natural but non-self-similar form. This trajectory defined what we called the Shadow Zone in analogy with optics. If one thought of the subducted fluid as ventilating the lower thermocline, this was the region not reached by any ventilated pathway. It remained in the *shadow* of the eastern boundary and was not illuminated by "rays" from the outcrop line and so did not carry potential vorticity information from the outcrop line. At this point the choice of the eastern region as stagnant was arbitrary and I will return to that point later. Continuing the solution with the Shadow Zone past the second outcrop line to the south led to a complex foliation of the solution and an increasing departure from the ideas that the thermocline could have a self-similar structure. Additional special domains appeared naturally in the model. An isolated region near the western boundary of the model was formed by a streamline in the subducted fluid that emanated from the intersection of the western boundary and the outcrop line. It seemed unlikely to us that such a zone would be at rest and with the Rhines and Young ideas in mind we entirely arbitrarily assigned that zone a constant potential vorticity equal to the potential vorticity on the bounding streamline. The idea was simply that eddy mixing and recirculation in this region might result in homogenized potential vorticity.

An earlier issue that arose in conversations with Hank was to what part of the subtropical gyre the model should be thought to apply. Hank's suggestion was that we should think about a so-called "mid-ocean gyre" one that had a rather artificial distribution of Ekman pumping whose integral in longitude vanishes at each latitude so that as much fluid downwelled as upwelled on each latitude circle. This would allow a circulation that would close without the need of western boundary currents that were not in our model. We talked a great deal about the advantages of this type of model but I believed strongly that the effect of our theory on the community would be greater if we just boldly presented it as a theory for the entire subtropical gyre and accepted that some undetermined, western part of the solution would be "contaminated" by effects from an active western boundary layer not considered in our model. We also attempted to extend the solution into the subpolar region but it was clear to us that our solution was an arbitrary one, deficient physically even if formally correct.

Christmas of 1981 was approaching and I took upon myself the task of preparing a manuscript describing our theory to which Jim added a good section describing what observational evidence could be mustered in its support. I wrote the paper at high speed, longhand in those days, and my consistent misspelling of the word "ventilated" (I was overly generous with the number of *l*'s) was a source of amusement to both Hank and Jim. I mischievously included an allusion to the *Communist Manifesto*, so subtle that no one has ever remarked on it to me.[5] A sometime tradition at Woods Hole was the Christmas GFD seminar in which serious work was presented with a humorous touch. Melvin Stern once gave it dressed as Santa. We were eager to present our work for the first time in that forum. Nancy Copley prepared a wonderfully witty announcement (see the Christmas Appendix) and Jim and I presented the observational and theoretical parts of the work with Hank squirming nervously in the audience. To finish off the seminar I had borrowed from a local theater company that was putting on a Christmas performance of Amahl and the Night Visitors three kingly crowns and rewrote the lyrics to a well-known carol called, for the seminar, "We Three Kings of Geostrophy Are" (see the Appendix) and to our amazement we succeeded in persuading Hank, always reluctant to perform in public, to join us in the singing. Fortunately, a member of the audience had a camera and the amazing moment is captured for posterity and is shown in the Appendix.

We were in high spirits. While I was drafting the paper Hank would drop by from time to time to discuss the work and the paper and once remarked with quiet satisfaction, "No one will ever think about the thermocline the same way again." It was not a boast but a sense of serious accomplishment he expressed and the three of us felt very close. We submitted the paper and naturally obtained lukewarm reviews but the paper was nevertheless accepted for publication. When the galley proofs were sent to me for correction I remember working on them one Saturday night and feeling very content with the paper and the work. The contentment was of short duration. Early the following morning the phone rang and it was Hank. "Have you seen Killworth's article in *Ocean Modelling?*" My reply was "no" and Hank said that although our paper was not yet published it was already being criticized but he was not sure of the basis. He said he would come right over and when I invited him in for a cup of coffee he nervously thrust the issue of *O.M.* into my hands and said, "Here, you take care of this" and rushed away. The substance of the article's argument was that any ideal-fluid thermocline model that allowed a nonzero stratification on the eastern boundary had to be entirely at rest to be consistent with the adiabatic, geostrophic equations. Our ventilated model with a nonzero thickness of the lowest layer on the eastern boundary was, therefore, somehow incorrect if it allowed any motion at all. The result was based on an ingenious theorem of Killworth's that used a Taylor series expansion of all variables from the eastern boundary and the equations to prove that all velocities were zero on the eastern boundary unless the stratification was zero, and

[5] "A . . . conceptual difficulty . . . haunts the history of the thermocline problem."

further, that all the higher derivatives forming the Taylor series would also have to be zero. Hence, any correct solution would be a null one and not at all like ours. After a few really bad minutes I called Hank and pointed out that the theorem would be valid in a region near the eastern boundary but only up to the shadow zone boundary where the first and higher derivatives of the interface thicknesses were not continuous. The bounding streamline for the Shadow Zone is a characteristic of the hyperbolic advective system and weak singularities in the derivatives of the variables are natural and limit the region of validity of the Taylor series . What Killworth's theorem actually proved was that our Shadow Zone needed to be truly at rest and so we had chosen well that solution.

Of course, in spite of our overall satisfaction we also realized that our analysis in many ways fell short of being a full solution to the thermocline problem. We did not explain what determined the layer thickness on the eastern boundary (later work allowed many layers to have such finite thicknesses, e.g., Pedlosky and Robbins, 1991), we still had to find the connection between our model and that of Rhines and Young (that came later with a joint paper, Pedlosky and Young, 1983), and we were keenly aware of the very provisional nature of the constant potential vorticity solution we had desperately employed in the nonventilated western "Pool" region of our model. Several important theoretical efforts have since been made to discuss that question, alter our solution, and put that part of the theory on firmer ground, in particular the recent interesting work of Dewar *et al.* (2005) and Radko and Marshall (2005). The biggest lack, still valid in my opinion, is the lack of coupling with the thermohaline circulation. Some progress along these lines followed the work of Salmon (1990) and Samelson and Vallis (1997) that carried through the original, heuristic argument of Welander (1971) and demonstrated the presence of a region of sharp gradients at the base of the adiabatic thermocline with the associated prediction of a cross isopycnal velocity that links the thermocline and the abyssal circulation. Additionally, Huang (1988, 1989) has developed a numerical implementation of a multilayer model whose limiting form as the layer number increases is equivalent to a continuously stratified model of the thermocline. Extending the theory into the equatorial zone also provides a simple model for the Equatorial Undercurrent and the shape of the equatorial thermocline and emphasizes a link between the midlatitude and equatorial oceans (e.g., Pedlosky, 1987).

With all its flaws, the ventilated thermocline remains a useful null hypothesis. We thought of it as an attempt to see just how much of the thermocline could be understood ignoring all other physical processes than just the simplest adiabatic, laminar dynamics. Eddy dynamics, micro-scale diffusion, air–sea buoyancy fluxes were all ignored on aesthetic rather than deductive grounds. At least some of these have since been proven to have been good guesses (see Ledwell *et al.,* 1993); others are probably less so. Indeed, it is so simple a solution that perhaps its greatest value is to serve as a tool for further improvements, alterations, or even, should it prove to be necessary, a clear basis for considering radical alternatives. After all, the rule is: what is modern today will seem primitive and dated in the future.

APPENDIX: THE 1981 CHRISTMAS GFD SEMINAR

The first seminar in which we discussed the theory of the ventilated thermocline was in Woods Hole on December 18, 1981. The announcement, prepared by Nancy Copley, is shown below.

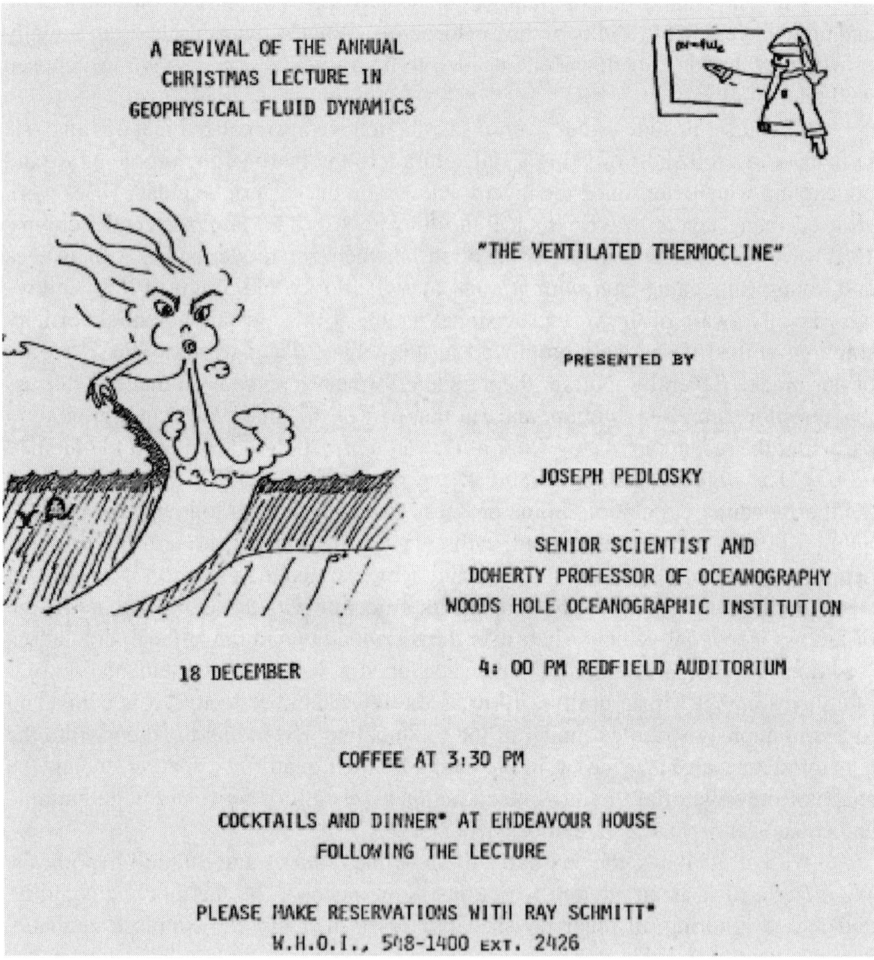

A REVIVAL OF THE ANNUAL
CHRISTMAS LECTURE IN
GEOPHYSICAL FLUID DYNAMICS

"THE VENTILATED THERMOCLINE"

PRESENTED BY

JOSEPH PEDLOSKY

SENIOR SCIENTIST AND
DOHERTY PROFESSOR OF OCEANOGRAPHY
WOODS HOLE OCEANOGRAPHIC INSTITUTION

18 DECEMBER

4: 00 PM REDFIELD AUDITORIUM

COFFEE AT 3:30 PM

COCKTAILS AND DINNER* AT ENDEAVOUR HOUSE
FOLLOWING THE LECTURE

PLEASE MAKE RESERVATIONS WITH RAY SCHMITT*
W.H.O.I., 548-1400 EXT. 2426

After the formal seminar, Hank Stommel, Jim Luyten, and I sang the following reworked carol, "We Three Kings of Geostrophy Are."

We three kings of Geostrophy are
Tracing streamlines that traverse afar,
The wind-driven gyre (but not quite entire),
Following yonder isobar.
(Chorus)

Oh, Oh, isobar of ideal lubricity	(Joe)
Isobar of potential vorticity	(Jim)
Westward swerving, still conserving	(All)
Guide us to beta-helicity	(Hank)
	(All repeat).

REFERENCES

Charney, J. G., 1955. The Gulf Stream as an inertial boundary layer. *Proc. Natl. Acad. Sci. USA* **41**, 731–740.

Defant, A., 1961. *Physical Oceanography,* Vol. 1. Pergamon Press, Oxford. 729.

Dewar, W. K., R. M. Samelson, and G. K. Vallis, 2005. The ventilated pool: A model of subtropical mode water. *J. Phys. Oceanogr.***35**, 137–150.

Gill, A. E., 1985. An explicit solution of the linear thermocline equations. *Tellus* **37A**, 276–285.
Holland, W. R., and P. B. Rhines, 1980. An example of eddy-induced circulation. *J. Phys. Oceanogr.* **10**, 1010–1031.
Huang, R. X., 1988. On boundary value problems of the ideal-fluid thermocline . *J. Phys. Oeanogr.* **18**, 619–641.
Huang, R. X., 1989. On the three dimensional structure of the wind-driven circulation in the North Atlantic. *Dyn. Atmos. Oceans* **15**, 117–159.
Iselin, C. O'D., 1936. A study of the circulation of the western North Atlantic. *Papers in Physical Oceanography and Meteorology.* Massachusetts Institute of Technology and Woods Hole Oceanographic Institution, Vol. IV, 101 pp.
Iselin, C. O'D., 1939. The influence of vertical and lateral turbulence on the characteristics of the waters at mid-depths. *Trans. Am. Geophys. Union* **20**, 414–417.
Ledwell, J. A., J. Watson, and C. S. Law, 1993. Evidence for slow mixing across the pycnocline from an open-ocean tracer release experiment. *Nature* **364**, 701–703.
Luyten J. R., J. Pedlosky, and H. Stommel, 1983. The ventilated thermocline. *J. Phys. Oceanogr.* **13**, 292–309.
Montgomery, R. B., 1938. Circulation in upper layers of southern North Atlantic deduced with use of isentropic analysis. *Papers in Physical Oceanography and Meteorology.* Massachusetts Institute of Technology and Woods Hole Oceanographic Institution, Vol. VI, 55 pp.
Needler, G. T., 1967. A model for the thermohaline circulation in an ocean of finite depth. *J. Mar. Res.* **25**, 329–342.
Pedlosky, J., 1969. Linear theory of the circulation in a stratified ocean. *J. Fluid Mech.* **35**, 185–205.
Pedlosky, J., 1987. An inertial model of the equatorial undercurrent. *J. Phys. Oecanogr.* **17**, 1978–1985.
Pedlosky, J., and P. Robbins, 1991. The role of finite mixed-layer thickness in the structure of the ventilated thermocline. *J. Phys. Oceanogr.* **21**, 1018–1031.
Pedlosky, J., and W. R. Young, 1983. Ventilation, potential vorticity homogenization and structure of the ocean circulation. *J. Phys. Oceanogr.***13**, 2020–2037.
Radko, T., and J. Marshall, 2005. The leaky thermocline. *J. Phys. Oceanogr.* **34**, 1648–1662.
Rhines, P. B., and W. R. Young, 1982a. A theory of the wind-driven circulation. I. Mid-ocean gyres. *J. Mar. Res.* **40** (Suppl.), 559–596.
Rhines, P. B., and W. R. Young, 1982b. Homogenization of potential vorticity in planetary gyres. *J. Fluid Mech.* **122**, 347–367.
Robinson, A., and H. Stommel, 1959. The oceanic thermocline and the associated thermohaline circulation. *Tellus* **XI**, 295–308.
Rossby, C. G., 1937. Isentropic analysis. *Bull. Am. Meteorol. Soc.* **18**, 201–209.
Salmon, R., 1990. The thermocline as an "internal boundary layer". *J. Marine Res.* **48**, 437–469.
Samelson, R. M., and G. K. Vallis, 1997. Large scale circulation with small diapycnal diffusion: The two-thermocline limit. *J. Marine Res.* **55**, 223–275.
Stommel, H., 1958. The abyssal circulation. *Deep-Sea Res.* **5**, 80–82.
Stommel, H., and G. Veronis, 1957. Steady convective motion in a horizontal layer of fluid heated uniformly from above and cooled non-uniformly from below. *Tellus* **IX**, 401–407.
Sverdrup, H. U., M. W. Johnson, and R. H. Fleming, 1942 *The Oceans: Their Physics, Chemistry and General Biology*. Prentice–Hall, Englewood Cliffs, NJ, 1077 pp.
Veronis, G., 1969. On theoretical models of the thermohaline circulation. *Deep Sea Res.* (Suppl.) **16**, 301–323.
Welander, P., 1959. An advective model of the ocean thermocline. *Tellus* **XI**, 309–318.
Welander, P., 1971a. Some exact solutions to the equations describing an ideal-fluid thermocline. *J. Mar. Res.* **29**, 60–68.
Welander, P., 1971b. The thermocline problem. *Philos. Trans. R. Soc. London Ser. A* **270**, 415–421.

Chapter 10

Sextant to Satellite

The Education of a Land-Based Oceanographer

S. GEORGE PHILANDER

The second half of the twentieth century was a most remarkable period in the field of physical oceanography. Into the 1980s oceanographers were still relying on stars and sextants to gather information as vitally important as the position of a ship at sea. Asking for "a tall ship and a star to steer her by" was not merely a romantic wish, but a practical necessity. Today it is possible to pinpoint the location of a ship by means of the Global Positioning System (GPS) which depends on several satellites in space. Such a remarkable change presumably had a profound effect on the way oceanographers see themselves, and the way they relate to the oceans. Sextants are the instruments of explorers—individualists who brave the elements and venture into uncharted territory in search of new worlds. Did the launching of the Sputnik, the first artificial satellite, in 1957, cause oceanography to lose its romance? Are there any worlds still to be conquered?

No, oceanography has not lost its romance, and, yes, there still are exciting discoveries to be made. I did not anticipate these answers when, during the 1990s, I tried to fathom why I was becoming more and more discontented with developments in my field, physical oceanography. Was my disquietude simply part of growing old, as

---

S. George Philander, Professor, Princeton University.

several of my contemporaries, including Dennis Moore, Ed Sarachik, Jacques Merle, and Jay McCreary, suggested? Or are there objective reasons for being concerned about the activities of oceanographers today?

There are ample reasons for nostalgia about the recent, post-Sputnik past. The launching of that satellite changed oceanography radically, but not adversely; it proved a bonanza for science in general, physical oceanography in particular. For a while scientists had easy access to resources for research, and enjoyed remarkable freedom to pursue science for the sake of science—science for the sake of understanding natural phenomena. The problems that oceanographers chose to investigate, related mostly to oceanic variability, yielded rich rewards. During the early 1990s it became evident that this exciting period was coming to an end. Three changes were clearly evident. (i) Oceanography, which used to be "bottom-up," with scientists identifying problems to be solved, then organizing and implementing programs to do so, had become "top-down" with managers and bureaucrats playing far more prominent roles. (ii) The field had lost some of its cohesion. For example, the perspectives of different groups on the same phenomenon, El Niño, became so different that the mere title of a paper on that topic now indicates whether the author is from Seattle, Los Angeles, or New York. (iii) Whereas the scientific and social interests of oceanographers used to be divorced—opposition to the war in Vietnam or apartheid in South Africa did not affect the choice of research topics—many scientists now want their science to influence policy.

At first I was unable to articulate why I was uncomfortable with these developments, especially the last one. By chance I came across two authors who provide valuable insights. One is Derek de Solla Price[1] who describes lucidly what happens when a "little science" grows into a "Big Science." (Although his is one of the most frequently cited books in the field of history and philosophy of science, surprisingly few scientists know of it.) The other author is the philosopher Isaiah Berlin[2] who explains eloquently why the profound differences between the worlds of science and of human affairs can cause the two worlds to be in conflict. A scientist who predicts El Niño, for example, should expect other scientists to be skeptical of the results; the practice of "organized skepticism" is essential for progress in science. On the other hand, a scientist who tries to persuade government officials that the prediction calls for the implementation of certain policies, will have no success if he is skeptical of the forecast. We should at least be aware of such dilemmas when we try to bridge science and politics. We should also appreciate that an excessive reliance on science in our attempts to deal with the societal aspects of environmental problems can be counterproductive if not disastrous. This is clearly evident in our attempts to cope with failures of the Indian monsoons early in the twentieth century.[3]

The continually changing social and political environment strongly influences the activities of scientists. The period from 1957 to 1982, a most unusual one for physical oceanography, was initiated and terminated by the same global political factors. The exciting period started because of concern that the former Soviet Union could be scientifically more advanced than the countries of the West. It ended when the fall of communism prompted the sponsors of scientific research to start demanding useful results. The political change did not occur until the late 1980s but in oceanography a change was evident as early as 1982 when an entirely unexpected, severe El Niño impressed on oceanographers their obligation to alert the public of imminent disasters.

Today oceanographers face a world very different from that of the 1960s. The competition for resources has become fierce, and the sponsors of research demand results that are of practical value. Maintaining a balance between the two complementary and equally important goals of science—science for the sake of understanding natural phenomena, and science for the sake of useful results—has become very difficult. Despite these differences, there are similarities between the situation today and that of the 1960s.

Then, an interest in oceanic variability led to the melding of scientists with different backgrounds; observationalists, theoreticians, oceanographers, and meteorologists all interacted closely, thus creating a new community. This integration led to beneficial changes in the manner oceanographers conduct their affairs. Once again, today, groups with different backgrounds need to merge in order to make progress on challenging and important scientific questions. Foremost is the need to anticipate how the current, rapid rise in the atmospheric concentration of carbon dioxide will affect the global climate. Changes in the oceanic circulation will be of central importance. Oceanographers have developed theories that explain that circulation, and the factors that can cause it to change, but a major problem is the lack of tests for the theories. The time scales of interest are so long, decades and more, that the measurements being made at present and in the near future will be of limited value. How do we respond to our sponsors who want answers soon?

Stringent tests for the theories are available in the geological records of the very different climates the Earth experienced in its distant past. Oceanographers are therefore obliged to take an active interest in paleoclimates. The challenge is daunting because physical oceanography has a vocabulary and methods—a culture—very different from that of paleoclimatology. The situation today has similarities with that of the 1960s; communities with different interests and methods need to cooperate. Familiarity with what happened over the past few decades can be a valuable guide. This article tells the story of those decades from the perspective of an unexceptional participant blessed with exceptionally good fortune. (For a more objective and detailed version of the same story[3] see Philander, 2004.)

UP TO 1957

Originally the oceans were studied mainly for the practical purpose of making journeys across the oceans shorter and safer. Studies of the ocean for its own sake—of conditions far below the surface for example—started when the *HMS Challenger* circumnavigated the globe between 1872 and 1876. Numerous similar expeditions, sponsored by various European colonial powers, followed immediately afterwards. The results were often published in atlases because, beneath its restless surface, the ocean was assumed to be unchanging. By the time of World War II the sum of all the measurements made on the various expeditions enabled oceanographers to arrive at a fairly accurate picture of the various components of the oceanic circulation. (The only major current to go unnoticed was the Equatorial Undercurrent which was discovered accidentally in 1953.) Efforts to explain how the winds determine different features of the circulation were off to a brilliant start with the studies of Sverdrup[4] and Stommel.[5] However, in 1954 Stommel[6] circulated a paper with the title "Why do our ideas about the ocean circulation have such a peculiarly dream-like quality?" Marrying measurements and theory would prove a major challenge.

1957 TO 1982

The launching of Sputnik contributed to a significant increase in the number of scientists entering the field of oceanography. I am one of those children of Sputnik. At Harvard University, under the tutelage of Allan Robinson, I was reared on Geophysical Fluids—stratified, rotating liquids. An interesting example of such a fluid is a cup of tea that has been stirred until it is in rigid body rotation. The tea returns to a state of rest in a surprisingly short time, significantly shorter than the diffusive time scale based on the dimensions of the cup. The explanation involves secondary circulations through thin diffusive boundary layers along the walls of the cup. In the 1960s, results such as these were applied to the oceanic circulation. My early contributions lengthened the list of "dream-like" ideas about the time-averaged oceanic circulation, but then, at MIT, Jule Charney encouraged me to explore oceanic variability.

Documentation of the time dependence of oceanic conditions requires simultaneous measurements over large areas, over extended periods. To acquire such data oceanographers were obliged to change their mode of operation significantly. Under the auspices of the International Decade of Ocean Exploration—the 1970s—they developed programs that coordinated the efforts of numerous investigators at many institutions in several countries. For guidance as regards appropriate measurements—the vastness of the oceans meant that only a few regions could be studied—many programs had "theoretical panels" whose members were mostly land-based and did not contribute significantly to the acquisition of data. When I joined the Geophysical Fluid Dynamics Laboratory (GFDL) in the early 1970s the director Joseph Smagorinsky wisely advised me to join a panel or two because they would provide access to the data that are needed to check theories and models. To me and my close collaborator Ron Pacanowski, GFDL was and continues to be the ideal institution for such involvement, not only because of its superb and supportive staff, but also because it has always attracted a steady stream of outstanding students and visiting scientists from whom I learnt a great deal.

I became the theoretician in a group of oceanographers, led by Walter Duing, who participated in a huge field program, involving an armada of ships, to study mainly atmospheric convection over the tropical Atlantic during the summer of 1975. At the time, so little was known about tropical oceanic variability that some of our colleagues questioned whether measurements over the relatively short period of a few weeks or months would reveal anything interesting. We consulted with Dennis Moore, but his ideas concerning strange waves that use the equator as a guide to race across ocean basins seemed so "dream-like" that we doubted their relevance to actual oceanic phenomena. We nonetheless decided to make most measurements very close to the equator where a curious undercurrent lurks. This led to a serendipitous discovery: energetic meanders of the equatorial currents with a period of a few weeks.[7] At about the same time, Richard Legeckis[8] discerned similar meanders in satellite photographs of sea surface temperature patterns in the tropical Pacific. A stability analysis of the time-averaged currents yielded results, concerning unstable waves, in accord with measurements of the spatial and temporal structure of the meanders.[12]

The relevance of theories concerning stable equatorial waves (developed by Taroh Matsuno[9] and Dennis Moore[10]) to actual oceanic phenomena remained in question until Carl Wunsch and Adrian Gill[11] used the available theories to explain curious oscillations, at periods of a few days, in sea level as measured at a number of islands in the tropical Pacific.

The oscillations in sea level at a few islands, and the meanders of the equatorial currents mentioned earlier, are phenomena of minor importance. Their discovery, and the success of theories[12] to explain their main features, nonetheless galvanized the community of tropical oceanographers because seemingly "dream-like" theories suddenly were of direct relevance to observations. Attention now turned to the response of the oceans to time-dependent winds.

How long, after an abrupt cessation of the winds, before a current such as the Gulf Stream disappears? George Veronis and Henry Stommel[13] first addressed this question and found that the answer depends on the time it takes Rossby waves to propagate across an ocean basin. In midlatitudes this is on the order of decades, but the speed of the waves increases with decreasing latitude and is infinite at the equator. Lighthill's seminal paper on the generation of the Somali Current,[14] which demonstrated that the removal of that singularity requires the introduction of equatorially trapped waves, motivated numerous theoretical and observational studies of a variety of phenomena. (Several people earned Ph.D. degrees correcting errors in Lighthill's paper. He neglected equatorial Kelvin waves, for example.) Henry Stommel persuaded younger colleagues to test these ideas concerning the generation of western boundary currents after the abrupt onset of the southwest monsoons. Bob Knox struck gold—time series of currents and winds—by asking officers at a British military base on the desolate island Gan within sight of the equator to make regular measurements from a small vessel. Ants Leetmaa spent much time in the Seychelles and east Africa, chartering vessels for additional oceanographic measurements.

In the Pacific, interest focused on El Niño. Klaus Wyrtki's analyses of tide-gauge records[15] revealed that the latter phenomenon is associated with a horizontal redistribution of warm surface waters across the tropical Pacific Ocean, induced by a relaxation of the trade winds. This presumably caused the equatorial currents to change radically. The French oceanographers in New Caledonia provided valuable data concerning the variability of currents in the western Pacific, but no information was available for the east. David Halpern[16] therefore developed instrumented moorings capable of measuring current and temperature fluctuations over prolonged periods and deployed those near the Galapagos Islands. The currents of the Atlantic Ocean are very similar to those of the Pacific, but the latter basin is significantly larger. One consequence is a dominance of the seasonal cycle in the Atlantic, of El Niño in the Pacific.

To interpret the empirical information concerning tropical variability, theoreticians developed a hierarchy of models. They range from relatively simple, highly idealized models, described by equations that can be solved analytically, to intermediate models, of the type developed by Cane[17] for simulation of the generation of the Equatorial Undercurrent, to the complex General Circulation Models (GCM) that require a supercomputer. The latter models are, in principle, capable of realistic simulations but, to interpret their results, and to analyze their deficiencies, require concepts and tools provided by more idealized models. The simple models can always be criticized for neglecting processes that in reality are important. Those models nonetheless are of great importance. For example, the concept of a Kelvin wave emerges, not from a GCM, but from a very simple analytical model.

For discussions of the rapid stream of new measurements and theories, Dennis Moore provided stimulating forums—informal, semiannual meetings where a small group of scientists engaged in lively debates. In the days before email, David Halpern edited a hugely successful newsletter that kept everyone abreast of new results concerning similarities and differences between the three tropical oceans. Progress was remarkably rapid. In the early 1970s very little was known about variability in tropical oceans; by 1982, oceanographers were able to document, explain, and simulate realistically the development and decay of the most intense El Niño in more than a century. The realism with which a GCM could reproduce the amplitude and timing of changes in the complex of equatorial currents, undercurrents, and countercurrents[18] persuaded several skeptics that the GCM is a powerful tool for studying the oceans. That this tool is widely used today is a tribute to Kirk Bryan and Michael Cox who pioneered its development.

El Niño of 1982 and 1983 was an opportunity for oceanographers to show how much they had advanced during the previous decade. However, the achievement was flawed in an important respect. Oceanographers failed to alert the public of the devastating El Niño of 1982 in a timely manner. They were enjoying such remarkable freedom to pursue science for the sake of science that they neglected the complementary goal of science—producing useful results. After 1982 they rapidly corrected this imbalance. Has the pendulum now swung as far in the opposite direction? Are we paying too little attention to science for the sake of science?

1982 TO THE PRESENT

After 1982, progress continued at a rapid pace. To stay informed about current conditions in the tropical Pacific, at and below the ocean surface, Stan Hayes[19] deployed the TAO array of instruments. To cope with the ever-increasing volume of data, Ants Leetmaa[20] started using an Oceanic General Circulation Model for operational purposes, each month producing maps of conditions in the Pacific, the oceanographic counterpart of the daily weather map. Having a seagoing oceanographer lead this activity contributed to the rapid acceptance of the GCM as a useful tool by a community which had been largely skeptical of the value of computer models of the ocean.

Up to 1982 oceanographers were interested mainly in the response of the oceans to changes in the winds. They attributed El Niño to an abrupt relaxation of the trade winds. But why do the winds relax? Meteorologists argue that the winds respond to changes in tropical sea surface temperatures, which in turn are induced by the winds. Bjerknes[21] first realized that this circular argument—the winds are both a cause and consequence of sea surface temperature changes—implies unstable interactions between the ocean and atmosphere. His paper gives a remarkably clear and accurate description of how El Niño develops, but the arguments are qualitative. Furthermore, Bjerknes was puzzled about the processes that terminate El Niño. Why is there a continual oscillation, the Southern Oscillation, which has El Niño as its warm phase? (In 1985 I introduced the name La Niña for the complementary cold phase.)

Some oceanographers continued to believe that El Niño is induced by a change in the winds (or westerly wind bursts). Others started investigating El Niño as one phase of a continual oscillation. Mark Cane and Steve Zebiak[22] first produced a coupled ocean–atmosphere model that spontaneously produced an interannual oscillation. The physical processes that maintain this mode were at first unclear until Paul Schopf and Max Suarez[23] proposed that the delayed response of the ocean to the winds, in contrast to the immediate response of the atmosphere to a change in sea surface temperatures, is of central importance. David Neelin,[24] David Battisti,[25] and their collaborators explored the continuous spectrum of natural modes that are possible; the delayed oscillator is but one member of the spectrum. The advances in our ability to observe and explain the tropical oceans and their interactions with the atmosphere were so rapid that, in June 1997, scientists were able to sound an alert that a major El Niño was developing, and could predict that California would experience a harsh winter six months later.

The launching of Sputnik brought constructive changes to the way oceanographers manage their activities. El Niño of 1982 brought further changes, but I found these ones disconcerting. I became aware of this during a small meeting in Miami in the early 1990s to plan the large, international program CLIVAR to study CLImate VARiability. Our efforts to define focused goals for this program were stalled when word spread through the room that help was on the way; an influential person happened to be in town and would be joining us soon. During my previous visits to Miami to plan programs—the GATE program for example—we often found ourselves in need of imaginative ideas to get a program moving. Usually a brilliant person would step in to provide a spark, a Jule Charney, or Henry Stommel, or Claes Rooth. On this occasion in early 1990, the person who stepped in was a "suit" from Washington who came to brief us on the type of program that politicians are likely to favor. How did nonscientists acquire the power to influence the activities of large numbers of scientists?

Informally organized activities, with scientists themselves doing the planning and implementation, are feasible when a science is "small." However, the enormous growth of physical oceanography over the past few decades, in the number of scientists, the number of papers they write, and the resources they need, transformed the field from a "small" into a "big" science. Such growth calls for formally structured activities that respond to demands from the sponsors of research for results that are of practical value. These can be salutary developments, provided we take care to distinguish between accomplishment and frantic activity—science is a highly undemocratic activity with a few people producing most of the important results[1]—and provided science for the sake of understanding natural phenomena continues to receive adequate support.

Growth sometimes leads to a loss of cohesion. That this was happening became evident in the special issue of the *Journal of Geophysical Research* (Vol. 103, June 29, 1998) devoted entirely to articles that review "Progress of El Niño Research and Prediction." Each article, written jointly by several distinguished scientists, is excellent, but is disappointingly disconnected from the other articles. The one that reviews oceanographic measurements makes essentially no reference to ideas presented in the paper on idealized models of El Niño. The articles on realistic models, and on prediction, are similarly detached, from each other and from the other articles in the special issue. When I mentioned this problem to colleagues, some denied that there is a problem, but others, David Neelin for example, agreed that different groups have very different perspectives on the same phenomena. In the summer of 2002 I called a meeting in Trieste, Italy, to address a seemingly straightforward question: given that El Niño corresponds to a natural mode of the coupled ocean–atmosphere, is that mode (a) strongly damped so that a continual oscillation is dependent on the presence of random disturbances? Or is it (b) self-sustaining and highly predictable? Or is it (c) sufficiently unstable to generate secondary instabilities so that, as in the case of weather forecasting, predictability is limited by errors in initial conditions? Each answer had ardent supporters!

After much discussion, a few people agreed that the available data sets are inadequate to distinguish between the three possibilities. The paucity of the data is evident in the way we have been surprised by each El Niño that has occurred thus far. Nobody anticipated that El Niño would be very prolonged in 1992, or that it would be exceptionally intense in 1997. As recently as 2004 some scientists sounded an alert that El Niño was developing, but withdrew it a few months later after nothing unusual had happened. As in the case of weather patterns which never repeat themselves, so each El Niño is distinct. We have not yet observed a sufficient number of realizations of El Niño to have rigorous tests for models that attempt predictions. Given that El Niño occurs every few years, enough data to test models and to settle debates about the predictability of El Niño should be available a few decades hence.

Recently the change in the properties of El Niño that occurred in the late 1970s has generated much interest in climate variability on time scales of decades and longer. This topic is related to the impact of future global warming on El Niño. Tentative results are available from climate models, but what are the tests for the models that will bolster confidence in their results? If the acquisition of data to test models that predict interannual El Niño episodes will take a few decades, how long before we can test theories for decadal variability?

NEXT

In science, measurements are essential for the testing of theories and models. The past few decades permitted the acquisition of extensive data sets only for phenomena with time scales shorter than a decade. That is why the biggest advances have concerned relatively brief phenomena that range from surface gravity waves to seasonal and interannual fluctuations such as El Niño. Now that our attention is shifting to phenomena with time scales much longer than a decade, we are obliged to take an interest in paleoclimates.

The remarkable achievements of paleoclimatologists, which include detailed descriptions of the astonishingly different climates our planet has experienced in the past, are not widely known among physical oceanographers. For example, everybody knows that dinosaurs were dominant during the extremely hot Cretaceous period, but few are aware of the strange journey that took our planet from that period to the present. The Earth experienced gradual global cooling during those 60 million years. The superimposed Milankovitch cycles were relatively modest at first but, around 3 million years ago, started to amplify, culminating in wild oscillations between prolonged Ice Ages and brief, warm interglacial periods such as the present one. At the moment, explanations for those phenomena have a decidedly "dreamlike" quality, thus presenting oceanographers with wonderful, new opportunities to explore strange worlds. To do so will require the cooperation of paleoclimatologists. As mentioned earlier, this is proving a challenge because of the different cultures of the two groups. Thus far, most attempts to overcome this problem have been "top-down" rather than "bottom-up." Important generals recruit officers who serve for a limited period of a few years on organizing committees that draft grand plans for large armies; the troops are expected to perform the same exercises. This approach is understandable, given that both paleoclimatology and oceanography are "big" sciences with huge communities and budgets. The results are impressively voluminous, filling many books and journals, but unfortunately those results continue to have a "dreamlike" quality.

Many of the problems we face today because of the rapid growth of science have been faced by our predecessors. Newton's contemporaries complained about too many people producing a "multiplicity of books; they doth so overcharge the world that it is not able to digest the abundance of idle matter that is everyday hatched and brought forth into the world." Those early scientists overcame the problem by forming "invisible" colleges, small groups of closely collaborating scientists not necessarily co-located. In the 1970s, "invisible" colleges took the form of small groups of scientists who organized programs to study phenomena such as coastal upwelling, and different facets of tropical oceanography in each of the three ocean basins. In the case of El Niño studies, a few key senior scientists provided strong encouragement, but credit for the accomplishments belongs to the junior scientists who operated without a commanding general. They, jointly, made the important contributions and orchestrated the marriage of what previously had been two separate disciplines, oceanography and meteorology. In the 1970s this "small science" approach enabled oceanography to progress from "dreamlike" to convincing, testable explanations for a variety of phenomena. A similar approach today will be a profitable investment.

Thirty years ago, a visionary NOAA administrator, Joe Fletcher, invited a small group of young scientists at different NOAA laboratories, all in their early thirties, to start a program to study a phenomenon, El Niño, about which very little was known at the time. Thus started my involvement in small programs that provided the most rewarding experiences, professionally and personally, of my career. The numerous meetings were not merely impersonal gatherings to discuss exciting measurements and theories, but were reunions of friends and their families in wonderful and exotic places. Today trips to Boston, Bologna, Tokyo, Paris, Venice, Kiel, Seattle, Los Angeles, Honolulu, and Reading are still for the purpose of attending meetings, and above all for the purpose of visiting dear friends and their families. Enterprising young scientists interested in forming small, "invisible" colleges to explore the strange worlds of the distant past can be assured that the rewards are wonderful. Older, established scientists and administrators should encourage and support them. Finding room for "small" science in our efforts to solve major problems is an imperative.

REFERENCES

1. de Solla Price, D. J., 1986. *Little Science, Big Science... and Beyond.* Columbia University Press, New York, 301 pp.
2. Berlin, I., 1991. *The Crooked Timber of Humanity.* Alfred A. Knopf, New York.
3. Philander, S. G., 2004. *Our Affair with El Niño; How We Transformed an Enchanting Peruvian Current into a Global Climate Hazard.* Princeton University Press, Princeton, NJ.
4. Sverdrup, H. U., 1947. Wind-driven currents in a baroclinic ocean with application to the equatorial currents in the eastern Pacific. *Proc. Natl. Acad. Sci. USA* **33**, 318–326.
5. Stommel, H., 1948. The westward intensification of wind-driven ocean currents. *Trans. Am. Geophys. Union* **29**, 202–206.
6. Stommel, H. M., 1995. Why do our ideas about the ocean circulation have such a peculiarly dreamlike quality? Or examples of types of observations that are badly needed to test oceanographic theories. In: *Collected Works of Henry M. Stommel.* American Meteorological Society, Boston.

7. Duing W., P. Hisard, E. Katz, J. Knauss, J. Meincke, L. Miller, K. Moroshkin, G. Philander, A. Rybnikov, K. Voigt, and R. Weisberg, 1975. Meanders and long waves in the equatorial Atlantic. *Nature* **257**, 280–284.
8. Legeckis, R., 1977. Long waves in the eastern equatorial Pacific. *Science* **197**, 1179–1181.
9. Matsuno, T., 1966. Quasi-geostrophic motion in equatorial areas. *J. Meteorol. Soc. Japan* **2**, 25–43.
10. Moore, D. W., 1969. Ph.D. thesis, Harvard University.
11. Wunsch, C., and A. E. Gill, 1976. Observations of equatorially trapped waves in Pacific sea level variations. *Deep-Sea Res.* **23**, 371–390.
12. Philander, S. G., 1976. Instabilities of zonal equatorial currents. *J. Geophys. Res.* **81,** 3725–3735.
13. Veronis, G., and H. Stommel, 1956. The action of variable wind-stresses on a stratified ocean. *J. Mar. Res.* **15**, 43–69.
14. Lighthill, M. J., 1969. Dynamic response of the Indian Ocean to the onset of the southwest monsoon. *Philos. Trans. R. Soc. London Ser. A* **265**, 45–93.
15. Wyrtki, K., 1973. Teleconnections in the equatorial Pacific Ocean. *Science* **180**, 66–68.
16. Halpern, D., 1987. Observations of annual and El Niño thermal and flow variations along the equator at 110W and 95W during 1980–1985. *J. Geophys. Res.* **92**, 8197–8212.
17. Cane, M. A., 1980. On the dynamics of equatorial currents with application to the Indian Ocean. *Deep-Sea Res.* **27A**, 525–544.
18. Philander, S. G., and A. Seigel, 1985. Simulation of El Niño of 1982–83. In: *Coupled Ocean–Atmosphere Models,* J. Nihoul, ed., Elsevier, Amsterdam, 517–541.
19. Hayes, S., L. Mangum, J. Picaut, A. Sumi, and K. Takeuchi, 1991. TOGA-TAO A moored array for real-time measurements in the tropical Pacific Ocean. *Bull. Am. Meteorol. Soc.* **72**, 339–347.
20. Leetmaa, A., and M. Ji, 1989. Operational hindcasting of the tropical Pacific. *Dyn. Atmos. Oceans* **13**, 465–490.
21. Bjerknes, J., 1969. Atmospheric teleconnections from the equatorial Pacific. *Mon. Weather Rev.* **97**, 163–172.
22. Zebiak, S. E., and M. A. Cane, 1987. A model El Niño–Southern Oscillation. *Mon. Weather Rev.* **115**, 2262–2278.
23. Schopf, P. S., and M. J. Suarez, 1988. Vacillations in a coupled ocean–atmosphere model. *J. Atmos. Sci.* **45**, 680–702.
24. Neelin, J. D., 1990. A hybrid coupled general circulation model for El Nino studies. *J. Atmos. Sci.* **47,** 674–693.
25. Battisti, D. S., and A. C. Hirst, 1989. Interannual variability in the tropical ocean–atmosphere system. *J. Atmos. Sci.* **46,** 1687–1712.

Chapter 11

Some Advances and Retreats in the Study of Ocean Circulation since 1935

JOSEPH L. REID

INTRODUCTION

This is an account of a few of the major findings about large-scale ocean circulation of the last few decades and perhaps some evaluations and comments on the past work.

We have collected vast amounts of data, from conventional hydrographic measurements with bottles and thermometers, bathythermographs, CTDs, ADCPs, current meters, drogues, drifters, and satellites. And we have computers that let us process, map, and calculate at speeds that we could not have dreamed of in 1950.

The data set has revealed many features new to the field, confirmed many others, corrected some serious errors of the past, and allowed many investigations that had not been possible before. Most of our findings about ocean circulation have resulted from measurements of the ocean. The major theoretical results so far—Ekman

Joseph L. Reid, Professor Emeritus, Scripps Institution of Oceanography.

transport, western intensification, and Sverdrup transport—followed from the results of measurements.

We have learned more about the manner of formation of the waters in the various basins, and their flow from basin to basin, the circulation of the large gyres, the flow along the boundaries, the penetration of the waters from the circumpolar flow into the various oceans, the equatorial circulation, and the transports of the major currents.

We have learned much of this from examining the patterns of the various tracers and from the density fields, which reflect the geostrophic flow.

We have also found that some of our early views of the circulation were wrong, and it is interesting to ponder why they were wrong and whether the earlier data could have given better answers.

DATA COLLECTION

Although a few large-scale expeditions had been carried out in the Atlantic Ocean (Fuglister, 1960) and in the Indian Ocean (Wyrtki, 1971), most work was proposed by individual investigators. One of these was the *Scorpio* expedition in March–July 1967 by Stommel, Reid, and Warren. It had two zonal lines from Australia to Chile. All three investigators were aboard on the first leg, along 43°S. It occurred to us that these were perhaps the longest such lines ever taken by a single ship with the same investigators, and the same equipment and reagents. This kept to the Aristotelian unities of place, time, and personnel, and our measurements should be internally consistent. At that time the geochemists were beginning to think about what could be learned if we were to measure the patterns of some of the isotopes such as carbon-14 and helium-3. Wally Broecker had one or two measurements of carbon-14 in the Atlantic and George Bien one or two in the Pacific. But more were certainly needed to make the data useful.

We thought it would be useful to equip a vessel to measure these and other isotopes along a line from the Aleutian Islands to Antarctica. Such a line would be a great help to the geochemists. We decided to suggest such an expedition for the consideration of the geochemists and the funding agencies, and that when Stommel got back after this cruise he should talk to them.

The funding agencies liked it and a program was proposed to measure carbon-14 and other substances in all three oceans. It was called GEOSECS (for geochemical sections) and was carried out from 1972 through 1978. More expeditions to assess the tracer fields were carried out in the 1980s in the Atlantic, the Transient Tracers in the Ocean (TTO), the North Atlantic Study (NAS), and the South Atlantic Ventilation Experiment (SAVE). Further contributions to the database continued and expanded up through the WOCE expeditions. Expeditions by various countries began to fill in the blanks in the ocean coverage. Recently there have been crossings of the Arctic Ocean.

FORMATION OF BOTTOM WATER

That the bottom water of the northwestern North Atlantic is formed by deep convection in the Irminger Sea had long been generally accepted. However, Lee and Ellett (1965) showed that the bottom waters of the Irminger Sea were not formed by overturn there. Their demonstration was by examination of Θ/S diagrams and did not have the immediate and broad impact that it should have. It was only when Worthington and Wright (1970) mapped the salinity on surfaces of constant potential temperature that the spills from the Norwegian and Greenland seas across the Greenland–Scotland sill, and their deep penetration, were generally recognized.

Why was it not discovered before? Data were adequate by the time of the *Meteor* expedition, or perhaps before to use the method of Worthington and Wright, as the displays of those data on maps at constant depths did not reveal the flow. Even Dietrich's (1969) maps of temperature and salinity at standard depths in the subarctic Atlantic do not themselves suggest any significant flow across the Greenland–Scotland Ridge except a shallow extension of the East Greenland Current, and no effect of it south of there at any depth. It is curious that this had not been settled much earlier. If Parr (1937) or Montgomery (1938) had carried out their studies on density surfaces farther north (near 50°N) instead of in the subtropics, the overflows from the Norwegian–Greenland Sea would have been obvious. Or if Montgomery and Pollak (1942) had not only traced the depth of some sigma-t surfaces from the *Meteor* Atlas, but also traced the salinity on those surfaces, the overflow would have been seen.

THE ANTICYCLONIC GYRES

Within the subtropical zone of each ocean there is a large anticyclonic gyre. Near the sea surface the westward limb of the gyre lies along 10° to 15° latitude, but below 500 m it lies poleward of 20° latitude. The great anticyclonic gyres contract poleward beneath the surface. This had first appeared in the *Meteor* Atlas. Wüst's (1935) study of the Atlantic found the tongue of high salinity from the Mediterranean outflow to be flowing westward from the Straits of Gibraltar all across the Atlantic. It joins the southern limb of the anticyclonic gyre, which flows westward directly beneath the southward-flowing surface water, and becomes part of the deep Gulf Stream. There is a hint of it in Iselin (1936). Defant (1941b) had mapped this feature in the North and South Atlantic, and Montgomery and Pollak (1942) had found it from the maps in the *Meteor* Atlas.

In the North Atlantic the contraction was also recognized clearly in Sverdrup *et al.* (1942). However, the maps of the Sverdrup transport prepared by Munk (1950) for the Pacific and Welander (1959) for the World Ocean showed only large anticyclonic gyres that resemble the near-surface circulation. The pattern of the deeper flow was obscured.

Was it the result of Munk's (1950) and Welander's (1959) total transport, which was dominated by the upper flow and showed no such contraction, that caused some investigators to try to get the salt tongue by special lateral mixing processes instead of advection(Richardson and Mooney,1975)?

Ivers (1975) suggested that the eastward limb of the North Atlantic anticyclonic gyre turned south and west and was joined by the highly saline outflow from the Mediterranean and accounts for the great tongue of high salinity extending westward from the Straits of Gibraltar. This flow corresponds closely to the maps of Montgomery and Pollak (1942) reproduced in Stommel's book on the Gulf Stream (1958).

Worthington (1976) could not resolve a large deep anticyclone with the salt tongue and simply restricted the gyre to the northwestern North Atlantic, north of the salt tongue.

This deep poleward shift of the western limb of the great anticyclonic gyre is also seen in the South Atlantic, South Indian, and North and South Pacific oceans. Coats (1981) and Young and Rhines (1982) proposed some explanations for it.

THE DEEP FLOW ALONG THE WESTERN BOUNDARY

It was known quite early that beginning near Cape Hatteras there is a deep northward flow beneath the shallow flow from the Florida Current but its source was not known. Iselin (1936) had proposed that the deep Gulf Stream observed north of the Blake Plateau must come in part from a part of the Stream which curves back to the right after passing the Grand Banks, and moves southwestward. He suggested that other sources might be an Antilles Current just north of the West Indies or a trade wind current south of it. However, the deep western boundary current flows southward along the Indies and flow near the Antilles does not contribute to the deep Gulf Stream.

Wüst's (1935) maps of the Upper and Middle North Atlantic Deep Water were based on the core method, vertical maxima in salinity and in oxygen, which extended southward, and thus did not directly show any evidence of the deep Gulf Stream, but did show a deep southward flow near or beneath the Gulf Stream, that extended near the western boundary from about 55°N to about 55°S.

His presentation of the deep southward flow was cited by Dietrich (1936), Rossby (1936), and Iselin (1936). Dietrich had noted a minimum layer of oxygen just off Chesapeake Bay near 800 m that extended more than 800 km offshore and proposed that it might represent a layer that was moving only very slowly, perhaps a "level of no motion." He showed that the geostrophic shear would accommodate a northward flow above 800 m with a southward flow below, though his calculation extended only to 1000 m. He cited Wüst's (1935) proposed deep southward flow in support.

Rossby (1936) referred to Dietrich's (1936) oxygen minimum layer and his assumption that it must be a layer at rest. But he wrote that "the assumption that the oxygen layer may be regarded as a zero surface for the velocity distribution by necessity implies that there is a fairly strong motion upstream, below the minimum layer" (p. 39). He extended Dietrich's calculations below 1000 m and he showed that using top-to-bottom data on the Chesapeake Bay section with a zero flow near 800 m would imply "that the amount flowing upstream must be in excess of 78 $\times$ $10^6 m^3$/sec, or more than double the 31 $\times$ $10^6 m^3$/sec carried downstream by the same calculation" (p. 40). He concludes that "Wüst's studies indicate that there must be a slow movement southward of the oxygen-minimum layer in the western half of the North Atlantic but there is no reason to assume that this motion is concentrated below the Gulf Stream as indicated by Dietrich" (p. 40).

Iselin (1936) suggested, on the basis of the average temperature distribution in a section between Chesapeake Bay and Bermuda, that this slow southward drift occurs considerably east of the Gulf Stream, in the vicinity of Bermuda. He was aware of these studies by Dietrich, Rossby, and Defant , but he held to a single deep Gulf Stream (his Fig. 27) whose sources had been uncertain. He concluded that the origin of this deep water was a turn back just south of the Grand Banks, leading to a southwestward flow offshore of the Gulf Stream, the Gulf Stream return flow. He stopped short of making the return connect to the Stream near Hatteras, but was close.

Swallow and Worthington (1961) used floats to measure the flow and find a reference surface near 1500 m. Transports calculated from the density field gave a deep southward flow of 6.7 $\times$ 10^6 m^3/sec beneath a northward flow of 56.4 $\times$ 10^6 m^3/sec.

Richardson (1985) combined surface drifters, SOFAR floats, and current meters along a section along 55°W of directly measured average zonal currents in and adjacent to the Gulf Stream. It showed the Stream with two flanking countercurrents, the southwestward flow along the boundary and a deep southwestward flow beyond the Stream. This was the sort of return flow Iselin had proposed, but much enlarged, and pictured by Sverdrup *et al.* (1942). Part of this southward flow turns northward near Cape Hatteras and becomes the deep Gulf Stream.

It is curious that Rossby and Iselin did not accept Wüst's deep southward flow near the Gulf Stream, which Dietrich had supported (though with a reference depth too shallow). Perhaps the greatest obstacle to acceptance of Wüst's proposal of a southward flow along the western boundary was that it did not show a deep Gulf Stream. Wüst could not find a core—a vertical extreme in salinity or oxygen—to define such a flow, so he could not map it, or perhaps did not believe in it.

CIRCUMPOLAR WATER AT MIDDEPTH IN THE NORTH ATLANTIC

The traditional view of the meridional flow of the deeper waters of the Atlantic was that all of the deeper flow was northward and the flow above it was southward. The boundary between the two flows was proposed to lie near the two-degree isotherm in potential temperature, which rises from near 4000 m in the North Atlantic to less than 3000 m in the South Atlantic. This was assuming that all of the water below was flowing northward and all of the water above was flowing southward, hence so many east–west sections were taken.

However, some circumpolar water passes northward at about 2000 m to 3000 m, east of the southward flow from the North Atlantic. The density range of the water from the circumpolar flow encompasses that of the North Atlantic, whose deep layer of high salinity, high oxygen, and low nutrients extends southward and penetrates the circumpolar water, separating it into upper and lower layers. The Circumpolar Water extends northward not only as Intermediate Water near 1000 m, as in Tsuchiya (1989), and along the bottom, but also a layer near 3000 m (Reid *et al.*, 1977). This layer from the south flows eastward and then northward and around the great gyre of the South Atlantic, northward across the equator in the east and northward just east of the southward flow along the western boundary.

Edmond and Anderson (1971) measured samples near 2500–3000 m just south of the equator that showed extrema in salinity, oxygen, and phosphate. These vertical minima in salinity and oxygen and maximum in phosphate indicate a flow from the Circumpolar Current, northward around the anticyclonic gyre and across the equator, east of the southward flow from the North Atlantic into which they interleave. The northward penetration of the salinity minimum near 2500 m are seen in Wüst's map of the Middle North Atlantic Deep Water, east of the southward flow, but he did not discuss them or put an arrow there to designate the sense of flow. The features can be seen near 3000 m on any north–south vertical section in the central Atlantic, but nearly every long straight north–south section that has been drawn in the Atlantic shows the penetration as deep blobs of higher or lower salinity, oxygen, and nutrients between about 40°S and 30°N. This is because the layer from the south does not flow only meridionally, but wends its way around and past the great gyres. It cuts across the meridional sections (Reid, 2005), spoiling the beautiful pattern that would be left from a simple meridional-only circulation.

(The referees of the Edmond and Anderson manuscript submitted to *Deep-Sea Research* did not permit the authors to claim any northward flow at middepth. It would have been heresy, so the indication of northward flow had to be omitted from the paper.)

GLOBAL BALANCE OF TRANSPORT

Several investigators, including Roemmich and McCallister (1989), McDonald (1998), and Talley *et al.* (2003), have carried out studies of large-scale circulation by balancing the total (top-to-bottom) geostrophic circulation along lines of stations enclosing an area. They find that balancing the total transport alone is not adequate: their solutions may balance the total but not the flow of each of the various layers. This appears to be the result of time variations. The lines of stations are not synoptic, and minor changes in speed over great distances and depth ranges can seriously alter transports in the various layers. That is, the middepth and deep oceans are also variable, and this includes large-scale flow. The time scales are fast enough to disrupt our simple schemes of steady-state geostrophic transport. The vertical shear is not constant over time, even in the deep waters.

ISOPYCNALS

Examination of patterns of characteristics along isopycnals that were first explored by Parr (1935) and Montgomery (1938) in the Atlantic has expanded since the 1950s and especially since the 1970s, when computers became available.

Ivers's (1975) study of the northern North Atlantic used what he called neutral surfaces. These have been developed further by McDougall (1987) and several other approaches have been suggested.

Reid (1994, 1997, 2003) has used the approximation suggested by Lynn and Reid (1968) in various large-scale studies in all three oceans. Large-scale investigations in all three oceans have involved some sort of isopycnal analysis.

Lozier *et al.* (1995) used patterns along isopycnal surfaces to prepare a climatologic atlas of the North Atlantic. With the vast amount of newer data and the interpolation along isopycnals it is much more complete than the earlier atlas of Levitus (1982), which averaged along isobaths, and gives a generally clearer picture. This and the study of Iorga and Lozier (1999a,b) show both the northward and westward extensions of the Mediterranean outflow.

Tsuchiya (1968) used isopycnals and relative geostrophic flow to map the upper circulation of the intertropical Pacific and (1989) the circulation of the Antarctic Intermediate Water in the North Atlantic.

Masuzawa (1972) used isopycnal patterns and relative geostrophic flow in a study of the North Pacific Ocean that dealt with the extension of the Kuroshio Current.

Mantyla and Reid (1995) displayed the deep tracers in the Indian Ocean and Reid (2003) added the adjusted geostrophic flow.

You *et al.* (2000) have made extensive use of the patterns of isopycnals in such studies as the North Pacific Intermediate Water and in the Atlantic–Indian Ocean exchange south of Africa (You *et al.*, 2003).

INTERMEDIATE WATER

Although the subsurface salinity minimum in the North Pacific Ocean had been apparent since the *Challenger* expedition, they had been supposed to derive from sinking of the low-salinity surface water of the far north (Uda, 1935; Sverdrup *et al.*, 1942). Later, Reid (1965) proposed that as water of the density of the salinity minimum had not been found at the sea surface in the North Pacific, the subsurface minimum there must occur by diffusion downward from the surface in the north and a subsurface lateral extension into and beneath the less dense but more saline surface waters of the central and eastern Pacific. The characteristics of the vertical minimum could be created and maintained over much of the subarctic gyre, where the surface salinity is very low.

In contrast, Talley (1991) has proposed that most of the characteristics of the water of the vertical minimum can originate in the Okhotsk Sea, directly from the surface, and that open-ocean mixing in the far north is less important. You *et al.* (2000) still hold for the downward mixing in both the Okhotsk Sea and the Gulf of Alaska. It remains an interesting problem.

In the Atlantic Ocean, Tsuchiya (1989) has traced the Antarctic Intermediate Water across the equator and along the western boundary and around the anticyclonic gyre up to 45°N.

There are similar subsurface minima in the South Pacific, South Atlantic, and South Indian oceans. The characteristics and density of the water at these salinity minima are found in the surface layer of the circumpolar flow, which is apparently the source of the subsurface salinity minima of lower latitudes (Taft, 1963).

GEOPOTENTIAL ANOMALY

One of the changes that have occurred during the last fifty years is that the mapping of geopotential anomaly, or dynamic height, to obtain relative or adjusted geostrophic flow, has become common for the Indian and Pacific oceans and may be beginning in the Atlantic.

Geopotential anomaly, or dynamic topography of the ocean was first proposed early in the twentieth century, but it was used by only a few investigators before 1950. This was partly because of the problem of a reference velocity.

I can imagine that in the early days of the field, after geostrophy had been introduced for studies of the atmospheric flow, someone, perhaps Bjerknes or Sandstrom, gave a lecture to introduce the geostrophic approximation to studies of the ocean. He would have explained that geostrophy could give only the vertical shear—the vertical difference in speed between two depths. To get the actual speed one needs something else—the flow measured or assumed at some depth. If, for example, it seemed that the flow changed sign at some depth, that would define a flow of zero, or a level of no motion.

There is really no reason to suppose that there is a surface of no motion everywhere, or that it is level or even continuous, but that such a distinguished scientist has used the phrase "level of no motion" as an example (even without arguing for it) gave it some standing. It has been used extensively in calculations of horizontal transports across the many zonal lines of stations in the Atlantic.

The relative geostrophic flow—the flow at some pressure with respect to a deeper pressure (500/2000 decibars for example)—has become more common. Over most of the ocean the flow is strongest near the surface and decreases rapidly below, and its direction may change. Using some depth as a reference for a shallow flow may approximate the actual flow, but it cannot give useful values of the transport, and using such a reference for flow at greater depths is likely to be very misleading in many cases.

The use of geopotential anomaly, or dynamic topography, began in the 1920s, with McEwen *et al.* (1930), Smith *et al.* (1937), and Koenuma (1937) in the Pacific, Helland-Hansen and Nansen (1927), Parr (1935), and Montgomery (1938) in the Atlantic, and Gordon *et al.* (1978) in the Antarctic. These all used some deep isobar as a reference. They recognized that the speeds they calculated were only relative and hoped for some solution.

Dietrich (1936) had proposed that the vertical oxygen minimum near 800 m beneath the Gulf Stream might indicate a depth of low or minimum flow in accord with Wüst's (1935) deep flow pattern. But Rossby (1936) showed that the vertical shear there, measured relative to 800 m, would have meant too much southward transport beneath it and too little northward transport above it. Did anyone accept a southward flow there? It seems to have been neglected by everyone except Defant (1941a,b).

After examining Dietrich's and Wüst's studies, Defant (1941b) proposed that there might be a reference surface that varied in depth from area to area to accommodate the different layers and sources. He also noted that though such a surface can be selected, it must be recognized as an assumption. He chose such a surface for the Atlantic Ocean and mapped the flow at depths from the sea surface to 2000 m. He referred the geostrophic shear to a surface of zero flow by proposing that a depth of no or minimum shear could be a depth of no flow, and thus a reference for calculating the speed. His reference surface was shallow near the equator (about 800 m) and sloped downward both north and south to about 2000 m. This, I believe, was to minimize the convergence in meridional flow.

But he made the reference surface slope up to the west near the western boundary. I believe that he did this because, after seeing Wüst's (1935) work and working with Wüst on the *Meteor* Atlas (Wüst and Defant, 1936), he knew that there is a deep southward flow beneath or near the Gulf Stream. The vertical shear there is monotonic, and a change in the sense of flow calls for a zero reference depth, in this case near 1500 m, with northward flow above and southward flow beneath. He could not extend his work below 2000 m because the data were scarcer there, and the fields were weaker, and would require greater accuracy in the measurements.

This choice of zero shear as zero flow did not receive any support. Not only was his method rejected, but also his results, and the whole concept of calculating geostrophic flow from the density field became suspect.

This is unfortunate, because his resulting "absolute" topography (Defant, 1941b) seems remarkably like the major features of the Atlantic flow that we recognize today. At the surface it shows the Gulf Stream and the Brazil Current, the two great anticyclonic gyres north and south of the equator, the North and South Equatorial currents and the Equatorial Countercurrent, and the subarctic gyre. It shows the narrowing of the anticyclonic gyres below 800 decibars, which had not been recognized, and the deep southward flow near the western boundary, the western boundary undercurrent, Wüst's "deep current."

This is not to say that his zero-shear choice really meant zero flow, and was a complete solution. Instead, the shear above and below his deep reference surface was strong enough to show something like the major circulation nearly everywhere. His results were not exact, of course, but much better than any that could be made from a fixed level of no motion and would have been a great step forward if they had been accepted.

It is surprising that this work was not taken more seriously. Swallow and Worthington (1961), in publishing the results of their own measurements of the southward flow, note that Defant had calculated velocities of 3 to 6 cm/sec at 2000 m, but it seems more likely that it was Stommel's (1957) survey of current theory that suggested their work.

After Defant (1941b) the next large-scale maps of dynamic topography were of the Pacific Ocean, made by Fleming *et al.* (1945) from the *Carnegie* expedition data. Although the measurements were sparse they gave a general picture of the anticyclonic gyre of the North Pacific.

Perhaps because Defant had not explained that his reference surface was really determined from a commonsense examination of the shear and the patterns of characteristics, not a questionable theory, dynamic topographies were not encouraged, though transports were, by Deacon (1937) and Sverdrup *et al.* (1942). A very few maps of geostrophic flow were produced for the Atlantic Ocean. Parr (1935) made some maps of dynamic topography in the Gulf of Mexico and Montgomery (1938) in the Atlantic. Stommel (1964) combined the maps of the dynamic topography of the Atlantic, Indian, and Pacific oceans by Defant (1941a), Lacombe (1951), and Reid (1961) to make a world map of the surface topography relative to 1000 decibars.

Dynamic topography has been used more extensively in the Pacific. Early examples are McEwen *et al.* (1930) from the United States and Canada, Uda (1935), Koenuma (1937), and Saito (1952) from Japan, Rotschi (1958) from France, and Wyrtki (1962) from Australia. It has been used extensively by various Soviet scientists. For example, Burkov has used it to study the circulation of the Pacific (1972) and later in a study of the World Ocean (1980) and Maksimov (1958) used it to study westward flow along the coast of Antarctica.

But it has been used much less extensively in the Atlantic. Is this a legacy of doubting Defant's results? Aside from the *Equalant* Atlas (Kolesnikov, 1973) almost the only Atlantic open-ocean map of dynamic topography after Defant that I can easily recall is that by Stommel *et al.* (1978). Perhaps this is the only map of steric height that ever came out of Woods Hole. (They would have fired anyone else who made one, but they couldn't fire Henry Stommel.)

RUMINATIONS

Hindsight reveals that we have made some mistakes. Some come from the assumption that something that has been generally accepted in the past must be correct, that it is an imponderable, something that one not only does not challenge, but can't even think about challenging. It has delayed some findings.

Why, for example, did the concept of overturn to the bottom in the Irminger Sea endure so long? There were data available that, used properly, could easily refute it, as was finally done by Lee and Ellett (1965).

Likewise, the spillovers from the Norwegian–Greenland Sea to the bottom of the North Atlantic would have been obvious if people had looked at tracers, not just on surfaces of constant depth, as in the atlases, which do not reveal directly the deep spills, but along other surfaces (Worthington and Wright on potential temperature, or even sigma-t, as Montgomery and Pollak did, but not far enough down).

Did Wüst's proposed southward flow fail to convince Rossby and Iselin simply because he did not show or refer to a deep Gulf Stream? His core method (the intermediate oxygen maximum) did not apply there, and could give no sense of northward flow. As both Rossby and Iselin knew the vertical shear there, they could have filled in the deep Gulf Stream by choosing a reference speed in the right place, and given us something like our present view. Why not? Defant did.

The apparent contradiction between a westward flow of the Mediterranean water, which had to cross the southward limb of the anticyclonic gyre, need not have arisen at all. Some people had been overwhelmed by the great anticyclonic gyre that Munk (1950) had found in the North Pacific and Welander (1959) found in the other oceans, and made the mistake of assuming that the gyres had the same shape all the way down, instead of contracting and leaving room for the westward flow. Sverdrup *et al.* (1942) had shown this shift much earlier.

The traditional view of the deep flow in the Atlantic, one deep layer flowing north along the bottom and another southward above it, led Wüst (1935) (and how many others?) to ignore the middepth northward flow that takes place east of his southward-flowing Middle Deep Water, and returns some of the water from the Antarctic Circumpolar Current northward toward the sources of the southward flow (Reid, 2005). Edmond and Anderson (1971) took data that showed it, though it was ignored at the time. The low oxygen and high nutrients from the south can be seen on meridional vertical sections only as separate blobs, because the northward flow meanders as it crosses the sections.

As computers continue to become faster and the information and the ideas we have about the circulation improve, better models will follow. But up to the present, most of what we have learned about the circulation of the ocean has come from the measurements that have been made. Are there any currents or large-scale patterns of flow in the ocean that were indicated by theory or models before they were revealed by observations?

With the recent emphasis on numerical modeling, some editors and referees have chosen to discount and discourage descriptive studies. This is Phariseeism. With the uncertainties of the constraints and other necessarily subjective inputs the models cannot yet claim eternal verity, but only numerical experiments and useful conjectures and tests.

Inverse models along ocean transects on a single line can balance transport in many ways, and provide various estimates of heat transport. But which is correct?

Some ancient logician has shown that the deductive method has no right to its starting point, and the inductive method no right to its conclusion. So much for logic. We need to measure, look, and test as well, using any means at hand.

We won't have a good estimate of the heat transport until we have the water going the right way.

REFERENCES

Burkov, V. A., 1972. *General Circulation of the Pacific Ocean Waters*. The Academy of Sciences of the U.S.S.R., P. P. Shirshov Institute of Oceanoglogy, Publishing House "Nauka," Moscow, 195 pp.

Burkov, V. A., 1980. *Obshchaya Tsirkulyatsiya Mirovogo Okeanna.* (General Circulation of the World Ocean). Translated from Russian. Published for the Division of Ocean Sciences, National Science Foundation, Washington, DC, by Amerind Publishing Co. New Delhi, 256 pp.

Coats, D. A., 1981. An estimate of absolute geostrophic velocity from the density field in the northeastern Pacific Ocean. *J. Geophys. Res.* **86**, 8031–8036.

Deacon, G. E. R., 1937. The hydrology of the Southern Ocean. *Discovery Reports* **15**, 1–124, pl. I–XLIV.

Defant, A., 1941a. Die relative Topographie einzelner Drückflachen im Atlantischen Ozean. Quantitative Untersuchungen zur Statik und Dynamik des Atlantischen Ozeans. *Wiss. Ergebn. Dtsch. Atl. Exped. "Meteor"* **6**(2) (4), 183–190, Beilagen X–XVIII.

Defant, A., 1941b. Die absolute Topographie des physikalischen Meeresniveaus und der Drückflachen sowie die Wasserbewegungen im Raum des Atlantischen Ozeans. Quantitative Untersuchungen zur Statik und Dynamik des Atlantischen Ozeans. *Wiss. Ergebn. Dtsch. Atl. Exped. "Meteor"* **6**(2) (5), 191–260, Beilagen XIX–XXIX.

Dietrich, G., 1936. Aufbau und Bewegung von Golstrom und Agulhasstrom, eine vergleichende Betrachtung. *Naturwissenschaften* **24**, 225–230.

Dietrich, G., 1969. Atlas of the Hydrography of the Northern Atlantic Ocean. Based on the Polar Front Survey of the International Geophysical Year, Winter and Summer 1958. Conseil International Pour l'Exploration de la Mer, Service Hydrographique, 140 pp.

Edmond, J. M., and G. C. Anderson, 1971. On the structure of the North Atlantic Deep Water. *Deep-Sea Res.* **18**, 127–133.

Fieux, M., R. Molcard, and A. G. Ilahude, 1996. Geostrophic transport of the Pacific-Indian Oceans throughflow. *J. Geophys. Res.* **101**, 12,421–12, 432.

Fine, R. A., K. A. Maillet, K. F. Sullivan, and D. Willey, 2001. Circulation and ventilation flux of the Pacific Ocean. *J. Geophys. Res.* **106**, 22,159–22, 178.

Fleming, J. A., H. U. Sverdrup, C. C. Ennis, S. L. Seaton, and W. C. Hendri, 1945. Observations and results in physical oceanography, graphical and tabular summaries, Carnegie Inst. Wash. Publ. 545, Scientific results of cruise VII of the Carnegie during 1928–1929 under command of Captain J. P. Ault, Oceanography-I-B, 315 pp.

Fuglister, F. C., 1960. Atlantic Ocean atlas of temperature and salinity profiles and data from the International Geophysical Year of 1957–1958. Woods Hole Oceanographic Institution Atlas Series, 1, 209 pp.

Gordon, A. L., E. Molinelli, and T. Baker, 1978. Large-scale relative dynamic topography of the Southern Ocean. *J. Geophys. Res.* **83**, 3023–3032.

Helland-Hansen, B., and F. Nansen, 1927. The Eastern North Atlantic. The cruises of the "Armauer Hansen", No. 1. *Geofys. Pub*l. **4**, 1–76, 71 plates.

Iorga, M. C., and M. S. Lozier, 1999a. Signatures of the Mediterranean outflow from a North Atlantic climatology, 1. Salinity and density fields. *J. Geophys. Res.* **104**, 25,905–26,009.

Iorga, M. C., and M. S. Lozier, 1999b. Signatures of the Mediterranean outflow from a North Atlantic climatology, 2. Diagnostic velocity fields. *J. Geophys. Res.* **104**, 26,011–26,030.

Iselin, C. O'D., 1936. A study of the circulation of the western North Atlantic. *Pap. Phys. Oceanogr. Met.*, **4**, 1–101.

Ivers, W. D., 1975. The Deep Circulation in the Northern North Atlantic, with Especial Reference to the Labrador Sea. Ph.D. dissertation, University of California, San Diego, 179 pp.

Koenuma, K., 1937. On the hydrography of the southwestern part of the North Pacific and the Kurosio. Part II. Characteristic water masses which are related to this region, and their mixtures, especially the water of the Kurosio. *Mem. Imp. Mar. Obs. Kobe* **6**, 349–414.

Kolesnikov, A. G., ed., 1973. *Equalant I & Equalant II—Oceanographic Atlas*. Vol. I. *Physical Oceanography.* UNESCO, Paris, 289 pp.

Lacombe, H., 1951. Application de la methode dynamique à la circulation dans l' Océan Indien, au printemps borèal, et dans l'ocen Antarctique, pendant l'ètè austral par. *Cah. Océanogr. Bull. D'Inform.* **3**, 459–479.

Lee, A., and D. Ellett, 1965. On the contribution of overflow water from the Norwegian Sea to the hydrographic structure of the North Atlantic Ocean. *Deep-Sea Res.* **12**, 129–142.

Levitus, S., 1982. Climatological Atlas of the World Ocean. NOAA Professional Paper 13, 173 pp.

Lozier, M. S., W. B. Owens, and R. G. Curry, 1995. The climatology of the North Atlantic. *Prog. Oceanogr.* **36**, 1–44.

Lynn, R. J., and J. L. Reid, 1968. Characteristics and circulation of deep and abyssal waters. *Deep-Sea Res.* **15**, 577–598.

Mantyla, A. W., and J. L. Reid, 1995. On the origins of deep and bottom waters of the Indian Ocean. *J. Geophys. Res.* **100**, 2417–2439.

Maksimov, I. V., 1958. A study of the Antarctic westward coastal current. (In Russian) *Inf. Byull. Sov. Antarkt. Eksped.* No. 2, 31–35. (Engl. Transl., *Sov. Antarct. Exped. Inf. Bull.* 2, 79–81, Elsevier, Amsterdam, 1964.)

Masuzawa, J., 1972. Water characteristics of the North Pacific central region. In: *Kuroshio*—Its *Physical Aspects,* H. Stommel and K. Yoshida, eds., University of Tokyo Press, Tokyo, 517 pp.

McDonald, A. M., 1998. The global ocean circulation: A hydrographic estimate and regional analysis. *Prog. Oceanogr.* **41**, 281–382.

McDougall, T. J., 1987. Neutral surfaces. *J. Phys. Oceanogr.* **17**, 1950–1964.

McEwen, G. F., T. G. Thompson, and R. Van Cleve, 1930. Hydrographic sections and calculated currents in the Gulf of Alaska, 1927 and 1928. *Rep. Int. Fish. Comm.* **4**, 1–36.

Montgomery, R. B., 1938. Circulation in the upper layers of southern North Atlantic deduced with use of isentropic analysis. *Pap. Phys. Oceanogr. Meteorol.* **6**(2), 55 pp.

Montgomery, R. B., and M. J. Pollak, 1942. Sigma-T surfaces in the Atlantic Ocean. *J. Mar. Res.* **5**, 20–27.

Munk, W. H., 1950. On the wind-driven ocean circulation. *J. Meteorol.* **7**, 79–93.

Parr, A. E., 1935. Report on hydrographic observations in the Gulf of Mexico and the adjacent straits made during the Yale oceanographic expedition on the Mabel Taylor in 1932. *Bull. Bingham Oceanogr. Collect.* **5**(1).

Parr, A. E., 1937–38. Isopycnic analysis of current flow by means of identifying properties. *J. Mar. Res.* **1**, 133–154.

Reid, J. L., Jr., 1961. On the geostrophic flow at the surface of the Pacific Ocean with respect to the 1000-decibar surface. *Tellus* **13**, 489–502.

Reid, J. L., Jr., 1965. Intermediate waters of the Pacific Ocean. *Johns Hopkins Oceanogr. Stud.* **2**, 85 pp, 32 figs.

Reid, J. L., 1994. On the total geostrophic circulation of the North Atlantic Ocean: Flow patterns, tracers, and transports. *Prog. Oceanogr.* **33**, 1–92.

Reid, J. L., 1997. On the total geostrophic circulation of the Pacific Ocean: Flow patterns, tracers, and transports. *Prog. Oceanogr.* **39**, 263–352.

Reid, J. L., 2003. On the total geostrophic circulation of the Indian Ocean: Flow patterns, tracers, and transports. *Prog. Oceanogr.* **56**, 137–186.

Reid, J. L., 2005. On the world-wide circulation of the deep water from the North Atlantic Ocean. *J. Mar. Res.* **63**, 187–201.

Reid, J. L., W. D. Nowlin, Jr., and W. C. Patzert, 1977. On the characteristics and circulation of the southwestern Atlantic Ocean. *J. Phys. Oceanogr.* **7**, 62–91.

Richardson, P. L., and K. Mooney, 1975. The Mediterranean Outflow—A simple advection–diffusion model. *J. Phys. Oceanogr.* **5**, 476–482.

Richardson, P. L., 1985. Average velocity and transport of the Gulf Stream near 55°W. *J. Mar. Res.* **43**, 83–111.

Roemmich, D., and T. McCallister, 1989. Large scale circulation of the North Pacific Ocean. *Prog. Oceanogr.* **22**, 171–204.

Rossby, C. -G., 1936. Dynamics of steady ocean currents in the light of experimental fluid mechanics. *Pap. Phys. Oceanogr. Meteor.* **5**, 1–43.

Rotschi, H., 1958. ORSOM III, croisière "Astrolabe", Océanographie Physiqued. Rapp. Sci. Inst. Franc. Océan., no. 8, 79 pp.

Saito, Y., 1952. On the Oyashio Current (Part I)—A synthetic research on the oceanographical state of the North Pacific Ocean. *J. Inst. Polytech. Osaka City Univ. Ser. B* **3**, 79–140.

Smith E. H., F. M. Soule, and O. Mosby, 1937. The Marion and General Greene Expeditions to Davis Strait and Labrador Sea. Scientific Results, Part 2—Physical Oceanography. U.S. Coast Guard Bull. No. 19, 259 pp.

Stommel, H., 1957. A survey of ocean current theory. *Deep-Sea Res.* **4**, 149–184.

Stommel, H., 1958. *The Gulf Stream: A Physical and Dynamical Description.* University of California Press and Cambridge University Press, 202 pp.

Stommel, H., 1964. Summary charts of the mean dynamic topography and current field at the surface of the ocean, and related functions of the mean wind-stress. In: *Studies of Oceanography*, K. Yoshida, ed. University of Washington Press, 53–58.

Stommel, H., P. Niiler, and D. Anati, 1978. Dynamic topography and recirculation of the North Atlantic. *J. Mar. Res.* **36**, 449–468.

Sverdrup, H. U., M. W. Johnson, and R. H. Fleming, 1942. *The Oceans: Their Physics, Chemistry, and General Biology.* Prentice–Hall, Englewood Cliffs, NJ, 1087 pp, 7 charts.

Swallow, J. C., and L. V. Worthington, 1961. An observation of a deep countercurrent in the western North Atlantic. *Deep-Sea Res*, **8**, 1–19.

Taft, B. A., 1963. Distribution of salinity and dissolved oxygen on surfaces of uniform potential specific volume in the South Atlantic, South Pacific, and Indian oceans. *J. Mar. Res.* **21**, 129–146.

Talley, L. D., 1991. An Okhotsk Sea water anomaly: Implications for ventilation in the North Pacific. *Deep-Sea Res.* **38**, S171–190.

Talley, L. D., J. L. Reid, and P. E. Robbins, 2003. Data-based meridional overturning streamfunctions for the global ocean. *J. Climate* **16**, 3213–3226.

Tsuchiya, M., 1968. *Upper Waters of the Intertropical Pacific Ocean.* Johns Hopkins Oceanographic Studies No. 4. Johns Hopkins Press, Baltimore, 85 pp, 32 Figs.

Tsuchiya, M., 1989. Circulation of the Antarctic Intermediate Water in the North Atlantic Ocean. *J. Mar. Res.* **47**, 747–755.

Uda, M., 1935. On the distribution, formation and movement of the dicho-thermal water to the northeast of Japan. *Uni to Sora* **15**, 445–452 (in Japanese).

Welander, P., 1959. On the vertically integrated mass transport in the oceans. In: *The Atmosphere and the Sea in Motion. Scientific Contributions to the Rossby Memorial Volume*, B. Bolin, ed., Rockefeller Institute Press, New York, 95–101.

Worthington, L. V., 1976. *On the North Atlantic Circulation.* Johns Hopkins Oceanographic Studies No. 6. Johns Hopkins University Press, Baltimore, 110 pp.

Worthington, L. V., and W. R. Wright, 1970. North Atlantic Ocean Atlas of potential temperature and salinity in the deep water including temperature, salinity and oxygen profiles from the Erika Dan cruise of 1962. Woods Hole Oceanographic Institution Atlas Series 2: 24 pp and 58 plates.

Wüst, G., 1935. Die Stratosphäre. *Wiss. Ergebn. Dtsch. Atl. Exped. "Meteor"* **6**, 109–288. Teil I: Schichtung und Zirkulation des Atlantischen Ozeans. (Also 1978, *The Stratosphere of the Atlantic Ocean. Scientific Results of the German Atlantic Expedition of the Research Vessel "Meteor" 1925–27*, Vol VI, Section 1, Berlin and Leipzig, 1935. (English translation edited by W. J. Emery) Amerind Publishing Co., New Delhi., 112 pp, Figs. 17–48, Inserts IX–XXXV.)

Wüst, G., 1937. Bodentemperatur und Bodenstrom in dep Pazifischen Tiefsee. *Veröff. Inst. Meeresk. Univ. Berl. N.F.* **35**, 1–56.

Wüst, G., and A. Defant, 1936. Atlas zur Schichtung und Zirkulation des Atlantischen Ozeans. Schnitte und Karten von Temperatur, Salzgehalt und Dichte. In *Wissenschaftliche Ergebnisse der Deutschen Atlantischen Expedition auf dem Forschungs—und Vermessungsschiff "Meteor" 1925–1927*, **6**, Atlas, 103 pl.

Wyrtki, K., 1962. Geopotential topographies and associated circulation in the western South Pacific Ocean. *Aust. J. Mar. Freshwater. Res.* **13**, 89–105.

Wyrtik, K., 1971. *Oceanographic Atlas of the International Indian Ocean Expedition.* National Science Foundation, Washington, DC.

You, Y., J. R. E. Lutjeharms, O. Boebel, and W. P. M. de Ruijter, 2003. Quantification of the inter-ocean exchange of intermediate water masses around southern Africa. *Deep-Sea Res.* **50**, 197–228.

You, Y., N. Suginohara, M. Fukasawa, I. Ysuda, I. Kaneko, Y. Yoritaka, and M. Kawamiya, 2000. Roles of the Okhotsk Sea and Gulf of Alaska in forming the North Pacific Intermediate Water. *J. Geophys. Res.* **105**, 3253–3280.

Young, W. R., and P. B. Rhines, 1982. A theory of the wind-driven circulation II. Gyres with western boundary layers. *J. Mar. Res.* **40**, 849–872.

Chapter 12

Towards the World Ocean Circulation Experiment and a Bit of Aftermath

CARL WUNSCH

INTRODUCTION

The World Ocean Circulation Experiment (WOCE) was the largest and most ambitious oceanographic experiment ever carried out. It was nearly 15 years in the planning, 10 years in execution, and the costs (depending upon what one counts) were of order U.S. one gigadollars spread over about 30 countries. Apart from the chapter by Thompson *et al.* (2001), comparatively little has been written about the origins of this unique program.

Here I will try to provide an informal, completely personal narrative of how WOCE came to be. I have read enough of the methods and concerns of professional historians to avoid making any claim that what is written here is any more than an anecdotal account, relying mainly upon my very imperfect memory, and incomplete records dating from 1977. I am looking back through the wrong end of the telescope.

Carl Wunsch, Professor, Massachusetts Institute of Technology.

Others who were involved from the beginning almost surely have a very different point of view. If a serious history of physical oceanography in the last quarter of the twentieth century is ever written, the material here should be regarded as at best a starting point.

At the outset, I note that much oceanography was conducted outside the WOCE framework, and it would be an error to claim that all of the advances that took place during the 1990s were attributable to it. But WOCE was surely the centerpiece, many observational and theoretical programs were put in place to take advantage of its existence, and the overlap of investigators working inside and outside the program was so great that attributing to WOCE much of the progress of that time is not a wild exaggeration.

ORIGINS

Background Science

One can trace the origins of what eventually came to be called WOCE to what, for a few of us, seemed to be an intellectual crisis in physical oceanography, circa 1975. In 1973, a field and theoretical program known as the Mid-Ocean Dynamics Experiment (MODE-1) had been carried out by a consortium of physical oceanographers from the United States and United Kingdom. This program, summarized by the MODE Group (1978), had exploited the then-new technologies of current meters, temperature recorders, bottom pressure sensors, XBTs, neutrally buoyant floats, and CTDs to demonstrate unequivocally the existence in the ocean of an intense eddy field. Prior to that time, the small-scale structures visible in hydrographic sections (e.g., Fuglister, 1960) had been regarded as a kind of fuzzy "noise" of no particular interest. A bit of information was available (e.g., Crease, 1962) suggesting from the primitive neutrally buoyant floats circa 1955, of remarkably intense, presumed transient motions at depth in the western North Atlantic. Fragmentary records existed from a number of comparatively brief measurements (see, for example, Monin *et al.*, 1977, Chapter 5). Physical oceanographers knew about internal waves, and were aware of the importance in the atmosphere of eddies [going back at least to Jeffreys (1933), and Victor Starr's work in the 1940s; see Lorenz (1967)]. But until about 1971, the technology simply did not exist to do more than speculate about the hypothetical importance in the ocean of time-dependent motions.[1]

[1] Soviet scientists had published a number of papers in their own literature that were interpreted by them as showing eddylike motions. Western scientists did not, however, take their results as seriously as they might have because of the language barrier; the primitive nature of the equipment (e.g., numbers were recorded by printing them onto a paper tape); and the locations such as the Black Sea. Monin *et al.* (1977, p. 133) list both Soviet and Western observations that, with hindsight, show eddies everywhere. (Note that in the Soviet literature, the term "synoptic" is used for the less proper Western adjective "mesoscale.")

Numerical modelling of the ocean had advanced greatly since the pioneering efforts of K. Bryan, G. Veronis, and a few others. But the slow, small computers of that era, combined with the very small deformation radius in the ocean conspired to prevent ocean models from being run in a high enough Reynolds number regime so as to become unsteady.

Between the limited observations and the sticky ocean models, the conventional picture of the ocean circulation was that of a laminar steady state. To this day, oceanographic textbooks still render the ocean circulation through pictures of large-scale scalar properties (e.g., temperature, salinity, oxygen) contoured and discussed as though the system is essentially steady and flowing only on the largest spatial scales—a geologist's view of the ocean. An analogy would be an atmospheric physics textbook that recognized only the mean, climate, state and failed to notice the presence of weather.

The results of MODE-1 and its troubled successor POLYMODE (see Collins and Heinmiller, 1989, for an account) showed that the ocean was likely an essentially turbulent fluid. Whether the turbulence had important dynamical and kinematic roles was unclear, but theory, and analogies with the atmosphere, suggested strongly than one could not simply assume it to be an annoying source of observational noise.

A parallel development, independent of oceanography, was the growing interest and concern about rising CO_2 levels. Several people, but notably Roger Revelle, were calling attention to the possibility of major climate change and insisting that the scientific community had to learn more about the implications. An indicator of the growing concern was the appointment of the so-called Charney Committee of the National Research Council in the summer of 1979 to examine the question. Three oceanographers (H. Stommel, D. J. Baker, and myself) were on the Committee, whose report (National Research Council, 1979) made a best-guess at the range into which global mean temperatures would be expected to rise. But a general theme of the brief report was the inability to be very definite about anything, particularly about inferences concerning the oceanic response, its uptake of carbon, and its thermal memory.[2]

Yet another relevant circumstance was the end of the so-called First GARP Global Experiment (FGGE), renamed, for the public, as the Global Weather Experiment. This program had been put together by the international meteorological community [Global Atmospheric Research Program (GARP)] to address the first of two overall goals—to improve weather forecasts. The organization of FGGE had left some

[2] Remarkably, the Charney Committee's estimate of the probable range for the expected increase in global mean temperature has hardly changed in the intervening decades. Much more is now known about the climate system than was true in 1979, and the continued agreement is largely fortuitous. Unhappily, some critics have interpreted this coincidence as implying that the ongoing scientific efforts to better understand climate change have been a waste of government money. This criticism is addressed by the Committee on Metrics for Global Change Research (2005).

of the oceanographic community feeling bruised, as the meteorological community wanted oceanographic ships as meteorological observing platforms, but cared nothing for the possible oceanography that might be done. With their much greater numbers, and national and international organizations, the weather forecasters essentially commandeered significant seagoing resources, leaving the oceanographers primarily as onlookers and passengers. But with the end of FGGE, GARP turned to their second goal—which was the understanding of climate change. When it came to climate, it was much harder to make a convincing argument that the ocean was largely irrelevant (although some meteorologists very seriously tried to do so, both then and today) and, internationally, efforts were begun to open a dialogue with the oceanographic community.

Thus, the situation in 1979 was that some oceanographers had a sense that the ocean was a far more dynamic place than historically believed; that it probably varied on all time scales—not just those of the newly discovered eddies; that we were being confronted with important societal questions about the ocean that were far beyond our ability to address, either theoretically or observationally. The question was what, if anything, could be done? If nothing could be done, it was clear that physical oceanography would become a marginal science of interest only to a few fluid-dynamics-oriented academics with the much larger meteorological community simply assuming that the ocean was basically passive ("swamp models" of the ocean are only now beginning to disappear). That NSF and ONR budgets for oceanography were shrinking was interpreted by some as demonstrating a field in decline, with no new ideas.

In 1979, I was invited to attend a meeting in Miami of a group called the Committee for Climate Change and the Ocean (CCCO) that had been formed by the IOC (Intergovernmental Oceanographic Commission) and SCOR (Scientific Committee for Ocean Research) and GARP to study the question of how one might address the problem of better understanding how the ocean influenced climate change. Thompson *et al.* (2001) describe the discussions that led to the calling of this meeting. I went, torn between the sense that we, as an oceanographic community, had to do something and that we probably could, and the realization that I was taking a tiger by the tail. If I was to be successful, I was condemning myself and others to years of organization and meetings.

To the extent that I can recall the thinking of the time, it was that our problem was primarily an observational one, and that sufficiently promising new technologies were being developed that, with some collective effort, might go a long way toward solving the fundamental problem. The observational problem was to (1) observe the ocean globally; (2) observe it spatially and temporally at sufficiently short intervals that one could define the dominant modes of variability everywhere. At a time when the main observational tool was still the ship, floats with tracking ranges of hundreds of kilometers, and expensive current meter moorings capable of operating for about a year, the question would immediately arise as to why anyone would think the global ocean could be adequately observed?

The technologies that I was aware of were several. CTDs were gradually becoming easier to use and more widespread. Autoanalyzers were available for nutrient measurements. Titration salinities had been replaced by conductivity methods. Transient tracers, tritium, helium-3, and chlorofluorocarbons were measurable. Bottom pressure gauges had become stable enough to yield months-long records. The neutrally buoyant float methods were rapidly advancing beyond the SOFAR method used in MODE-1 to RAFOS (Rossby *et al.*, 1986) and what eventually became the ALACE floats (Davis *et al.*, 1992). In the summer of 1977, Walter Munk and I (Munk and Wunsch, 1979) had stumbled on the idea of ocean acoustic tomography, which promised to provide large area integrals over the ocean. Perhaps most important, however, was the prospect of certain satellite measurements of the ocean, in particular scatterometry for winds, altimetry for circulation, and gravity for determining the absolute circulation.

Altimetry and tomography were my own particular foci, and as W. Munk describes the evolution of the acoustical capability elsewhere in this volume, perhaps I can be permitted some words about altimetry.[3]

I cannot do justice here even to the history of altimetry, much less all of the other technologies that were emerging at that time. I would argue, however, that altimetry has played a unique role as, to this day, it remains the only true global ocean measuring system (scatterometers and other devices measure parts of the forcing, not the ocean itself).

Altimetric Measurements

Like most physical oceanographers, I had no experience with remote sensing from space, when in 1974 I had a telephone call from Dr. Peter Bender, a space geodesist working for NOAA in Boulder. Peter explained that he was chairman of the Committee on Earth Sciences of the Space Science Board of the National Research Council, and that they were trying to write a report discussing, in part, what NASA should be doing to better understand the ocean. My response, which was a flat refusal, clearly startled Bender. I told him that NASA's contribution to oceanography seemed all hype—based upon a few not-very-accurate infrared measurements of sea surface temperature from space. Sea surface temperature was of much more interest to meteorologists than to oceanographers in any case, and I thought that NASA's public relations machinery was far outstripping the importance of its contribution. After a stunned silence on the other end of the telephone line, Bender said that if things were really so bad it was even more important that I should serve on the Committee, so that the Report would reflect the reality. In a weak moment, I then agreed.

At that time, NASA's oceanographic interests were focussed on the so-called SEASAT-A spacecraft which was to fly circa 1977. It is hard now to credit an era in

[3] In the end, tomography played only a small role in WOCE as the acoustic technology did not develop as rapidly as hoped. It may now be on the verge of large-scale use.

which NASA was looking for things to do. A committee of enthusiasts had been put together by NASA that proposed an ocean satellite to measure virtually everything from space that seemed technically possible, in some cases without much justification for what the measurement would say about the ocean. As part of the Bender Committee, I undertook to read the documentation justifying the decision that had already been made to fly SEASAT-A. (One question we were faced with was how a successor satellite—SEASAT-B—should be configured; it was taken for granted that there would be a follow-on of some sort.) The level of technical detail and justification for SEASAT-A in the reports would be regarded as extremely thin, bordering on the laughable, by today's standards. As I read through the documents, however, I finally came to the discussion of the altimeter that would be on the satellite. Although the report said little about how the measurements would be used, it became clear to me that if the instrument system could live up to the engineering specifications, it represented a very exciting possibility—the measurement of surface dynamic height from space at a useful level of accuracy. From the earliest days of the so-called dynamic method, about 1900, the direct determination of sea surface slopes relative to a reference surface (called the geoid) had been recognized as an important concept, but whose measurement was regarded as essentially impossible. Here was NASA explaining, in primarily engineering terminology, that perhaps it could be done. I got interested.

SEASAT (the "A" was dropped on launch) finally flew in 1978, but instead of running for several years, it failed after three months. (Rumors immediately circulated that it had been deliberately killed by the U.S. Air Force, who were supposed to have aimed a laser at it. In the aftermath of the Vietnam War, many scientists were deeply suspicious of the military, and there indeed had been great tension over whether the SEASAT measurements would be classified. The SEASAT saga remains to be written.) As it turned out, the failure after so short a time was something of a blessing. Cost overruns on the hardware and launch had eaten up the science analysis budget. With the failure, some money from the operations budget was made available to the science community to analyze what data there were. These proved adequate to show that the altimeter actually worked at the levels of accuracy and precision predicted by the engineers. For example, one could clearly see the Gulf Stream and associated rings (Wunsch and Gaposchkin, 1980; Cheney, 1982). The concept had been proven (see Figure 12.1).

A separate (long) paper would be required to describe the events that ultimately led to the launch of what is now known as TOPEX/POSEIDON, a U.S.–French mission that became the centerpiece of WOCE. Anyone who becomes involved with the formulation of a new mission will have their own stories of near-failure, bureaucratic and political craziness, heroic and not-so-heroic individuals, and plain luck. That TOPEX/POSEIDON was actually launched, and has performed far beyond its specifications for, as I write, almost 13 years (the agreed lifetime was 3–5 years) is in the nature of an engineering/scientific/political miracle that deserves its own history.

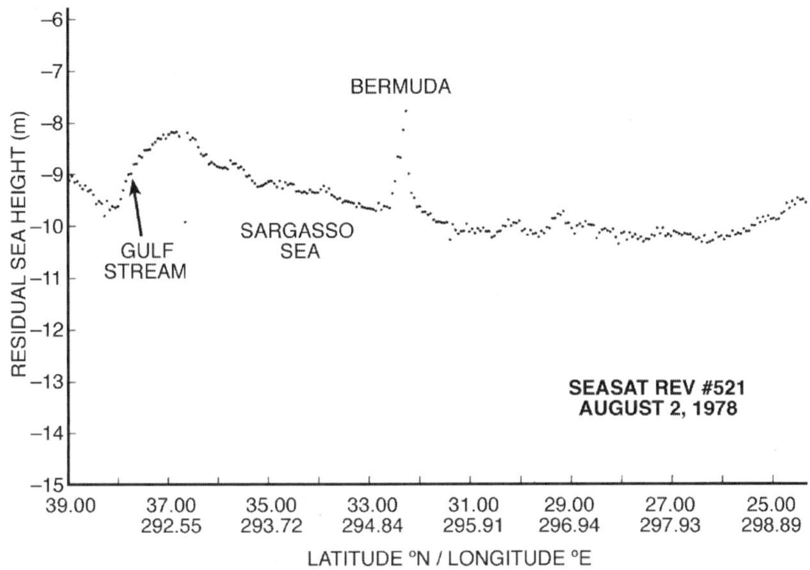

Figure 12.1. An early measurement (Cheney, 1982) from SEASAT showing the presence of the Gulf Stream in altimetric data. The presence of a Bermuda signal is evidence of the large geoid (gravity field) errors present in the data.

Modeling and Theory

By 1979, there were global coarse resolution numerical models, and small-scale, idealized geometry, eddy-resolving models. (See Figure 12.2, from Holland and Lin, 1975.) Moore's Law (Moore, 1965) was already widely known, and extrapolation of work already underway suggested that by about 1990 one would have the beginnings of global-scale eddy-resolving models.[4]

Anyone who understood models realized that the more sophisticated the model, the more demanding the requirements on the observations. It was obvious that numerical models of the ocean were about to outstrip any observational capability for testing them. There was a grave danger that the field would produce sophisticated, interesting models, without any ability to calibrate them. (This situation now exists in paleoclimate studies, where seemingly sophisticated models are compared to sparse, poorly understood observations.)

With a few rare exceptions, the coast-to-coast hydrographic surveys, epitomized by the Meteor surveys of the 1920s and the International Geophysical Year (IGY) surveys of the 1950s, had fallen from favor. They appeared to be of mainly qualitative

---

[4] The computer story involves much more than the number of circuits on a chip. Moore's Law is a metaphor for cheap storage, parallelization, input–output devices, and new software, that were required for the construction and use of models of a size and complexity far beyond what was possible in 1980.

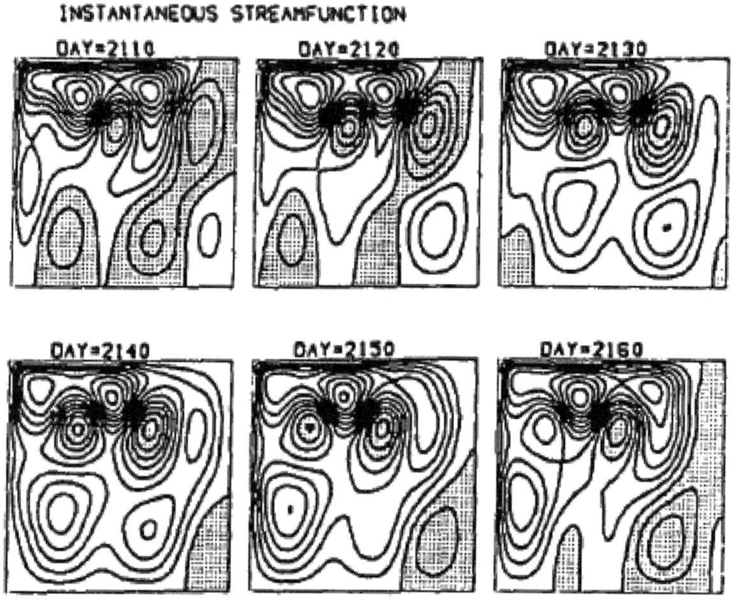

Figure 12.2. From Holland and Lin (1975) showing an early ocean model producing eddylike features. The model had one layer and was nominally 1000 km on a side.

use—and many, perhaps most, physical oceanographers had turned instead to the more scientific-seeming process studies of the era of the International Decade of Ocean Exploration (IDOE). These included MODE, focussing on the mesoscale variability, but also upwelling studies in various places, internal wave studies; the monsoon regime of the Indian Ocean; and so on. In contrast, observations of long hydrographic sections resulted primarily in atlas plates, quite beautiful, but more art than science, with the accompanying scientific papers being primarily descriptions of water masses, or unconvincing attempts to guess the absolute flow directions. By the middle 1970s, the notorious so-called level-of-no-motion problem, which had plagued oceanography from the earliest days of hydrographic surveys, was finally understood, and solved by inverse methods—in several guises (Wunsch, 1996). The advent of these methods meant that coast-to-coast hydrographic lines could be used quantitatively; it was also recognized that altimetry combined with an adequate gravity mission was an alternative method for determining the absolute flow field (Wunsch and Gaposchkin, 1980). With the new ability to calculate flow fields and transports without arbitrarily chosen levels-of-no-motion, it made sense to contemplate a proper "long-line" survey of the ocean.[5]

[5] Dean Roemmich and I (Roemmich and Wunsch, 1985) made the first trans-Atlantic hydrographic sections since the IGY (1958–59) during the summer of 1981. We had the use of inverse calculations specifically in mind, as well as the opportunity to see if the North Atlantic Ocean had changed measurably in the intervening years (it had, in a number of ways).

PROPOSING IT AND SELLING IT

In any event, with the sense that we could develop adequate technologies in a reasonably brief time period, that models would probably improve independent of any field program, and that we knew generally what needed to be done, I proposed at the Miami meeting that there should be an attempt to measure the ocean circulation and its variability, globally, as the oceanographic contribution to understanding the climate state. R. Stewart (Canada) made another specific suggestion: that it would be useful to attempt to formulate a complete, closed heat budget of the North Atlantic Ocean sector, including both atmosphere and ocean as a trial experiment for a possible later global one. Some combination of *in situ* observations of ocean and atmosphere, along with coupled models would be used to understand how heat was transported by both fluids, and how it was transferred between them. At some point, Stewart's proposal was labelled the "CAGE" experiment, as it would basically involve building a cage around the North Atlantic basin in both atmosphere and ocean. In response to the two proposals made at the Miami meeting, the CCCO appointed two Committees: one was chaired by Fred Dobson (Bedford Institute) to examine the prospects for CAGE; the other was chaired by Francis Bretherton (then Director of NCAR) to examine the prospects of a global experiment. The report of the CAGE committee (Dobson *et al.*, 1982) was very impressive and came to a startling conclusion—that CAGE was impractical, not because of the problems of observing the ocean, but because atmospheric measurements were inadequate to close the atmospheric side of the heat budget![6] This wholly unexpected conclusion effectively left the global ocean experiment alone as a serious proposal ("...the concept of a North Atlantic CAGE experiment lies battered and torn,..." from a letter of F. Dobson to the Committee on Climate Change and the Ocean, January 10, 1983).

Another, completely separate, program ultimately called TOGA (Tropical Ocean, Global Atmosphere) was being formulated and organized. TOGA has been described at length elsewhere (see Halpern, 1996, for a discussion of its origins). Suffice it to say that its flavor was very different, involving as it did a very large meteorological component, a goal of forecasting, and a hard insistence that only the upper few hundred meters of the near-equatorial ocean had to be understood in order to achieve its goals. The latter point of view, in particular, ultimately caused difficulties for what became WOCE.

The Bretherton committee, studying the option of a global ocean circulation program, eventually concluded that it might be feasible, and recommended that serious planning and study should begin.

That, of course, was when our real troubles started. The job was to organize something both nationally (the U.S. contribution was clearly going to be the dominant one) and internationally, on a scale never before tried by oceanographers,

[6] Much of the difficulty lay with the problem of calibrating radiosondes, whose offsets prevented the possibility of closing the atmospheric budget.

and without the managerial infrastructure available to the meteorologists who had organized FGGE—with their national meteorological agencies as a base. Oceanographers had nothing remotely resembling such governmental organizations.

PLANNING IT

Shortly after the CCCO discussion, and the appointment of the Bretherton Committee, I spent a year in Cambridge, England, with the help of a Guggenheim Fellowship. In addition, Walter Munk came for six months, and we split a Fulbright Award (inevitably then known to our wives as a half-bright award). During this period, when we were focussed on trying to turn ocean acoustic tomography into a practical observational method, I attended a Royal Society discussion meeting on oceanography in the 1990s for which Munk and I wrote a speculative paper (Munk and Wunsch, 1982) that laid out a rough vision of how the emerging technologies might be deployed to give a much more realistic understanding of the time-dependent ocean. [A less formal account appears in the Munk Festschrift (Garrett and Wunsch, 1984).]

How does one obtain legitimacy for a proposed national and international program? In the United States, recognition appeared to come through the National Research Council (National Academy of Sciences), through what is now called the Ocean Studies Board (OSB; the name has changed several times over the years. It was then called the Board on Ocean Science and Policy). A small self-appointed steering group (including Baker, Nowlin, Broecker, Wunsch) agreed to try to put together a U.S. national program. I went with some of the steering committee to a meeting of the Board in Washington where I presented the idea of a global ocean circulation program. That both Baker and I were members of the Board appeared to make the request particularly simple. To my very great surprise, the request was flatly refused. The Chairman of the Board (J. Steele, then Director of WHOI) announced that there would indeed be a national oceanographic program, but that it was to include biology, and he would be the chairman.

I returned from the OSB meeting convinced we had failed to even get out of the starting gate. About a week later, however, Steele telephoned me to say that, of course, we could have a workshop, and that the Board would endorse and help organize it. Someone had gotten to him in the interim. Steele was evidently fearful that the physical and chemical oceanographers would have a major program and that the biologists would be left out. Steele's efforts to construct a parallel biologically oriented program eventually became GLOBEC, but that is someone else's story.

A small steering committee (D. J. Baker, F. Bretherton, W. Broecker, J. McWilliams, W. Nowlin, F. Webster, and C. Wunsch) was appointed through the Ocean Climate Research Committee of the Board to organize a Workshop, which took place in August 1983 at the National Academy of Sciences building in Woods Hole, Massachusetts. About 70 people were officially present, including agency representatives and many from abroad. The resulting report (Ocean Climate Research Committee, 1984) was based upon various white papers plus discussion. Its publication was interpreted as endorsement of a U.S. program by the Academy, and by the U.S. government agencies which would have to fund it.

Internationally, the World Climate Research Program (WCRP, with headquarters in Geneva) through its own steering committee, was induced to appoint an international planning committee. The original committee membership was F. Bretherton, W. S. Broecker, J. Crease, K. F. Hasselmann, M. P. Lefebvre, A. Sarkysian, J. Woods, R. Kimura, and myself, as chairman. Because many of the results of WOCE bear directly on physical oceanographic problems, it is not widely recalled that WOCE was a climate experiment—and was accepted as such by the WCRP. Many oceanographic issues had to be resolved, but the goal was, and remained, to quantify the contribution of the ocean to control of the climate system, to provide a baseline against which future climate change could be measured, to understand the extent to which its variability existed, and what its consequences were.

There then proceeded to be several years of seemingly endless numbers of meetings (well over a hundred) devoted to determining (1) what we were trying to do and (2) how we would do it. Discussion meetings were focussed, variously, by technology, by ocean basin, and by scientific goal. A framework with two overall goals was produced (directed at producing data sets adequate to test the models expected circa 1990, and determining what kind of observation program would be adequate for indefinite monitoring of oceanic climate states, respectively).

A few events stand out. The initial WOCE planning envisaged including measurements and understanding of the ocean carbon uptake and redistribution problem, as the fate of fossil-fuel CO_2 was one of the driving uncertainties. It quickly became clear, both in U.S. national and international meetings, that the CO_2 problem could not be dealt with as an appendage to a program primarily in the hands of physical oceanographers. A major problem was that serious technical disagreements existed among the small community of people who measured oceanic CO_2 (e.g., C. D. Keeling, P. M. Brewer, and others) as to how it could be, or should be, done. Expertise necessary to distinguish between the competing arguments was not adequately represented on the steering committees. Furthermore, at least one member of the international committee (WMB) repeatedly insisted that WOCE should be a tracer measuring program alone, with discussion of altimetric satellites, conventional hydrography, and the like being a "dead end."[7]

---

[7] Letter from Broecker, 30 August 1988 to C. Wunsch. It is perhaps worth quoting from this letter as it demonstrates the divisions in the community over what needed to be done:

> "...the program is too much driven by satellite topography, rapid hydrographic sections and inverse modeling. In my view the approach is basically a dead end. The great hope of the future is atmospheric driven models.
>
> I agree that atmospheric driven ocean models must fit the temperature and salinity field (and that to some extent they currently fail this test.) However, one does not need a WOCE program to generate an observed temperature and salinity distribution. We have a perfectly adequate one for this purpose."

The letter was copied to 31 colleagues around the country, and was representative of several others in this vein, although more restrained than some.

It was finally concluded that a separate program, which became JGOFS, should be spun-off into the hands of the requisite experts, with a commitment (which was honored) for WOCE to provide shiptime and to generally collaborate. With hindsight, this decision was the right one, with WMB focussing his unhappiness primarily on the JGOFS organizers, not WOCE (but see Kerr, 1991).

Organizing national and international programs is a huge time sink. We took as the principle that coordination would be attempted only if was really required—because temporal simultaneity was essential. For example, important as modelling would be to WOCE, it did not require the same degree of international organization that the observational programs did. To a large degree, the modelling community was advancing with the growing computer power—a development that was out of the hands of oceanographers. They were already reasonably well-organized internationally, having periodic meetings that brought the main players together. A policy of "benign neglect" seemed to be appropriate, and seems to have worked reasonably well, although inevitably, some of that community chose to infer that WOCE was antimodelling. The most conspicuous WOCE modelling program was the community effort led by C. Böning, W. Holland, and others (the WOCE modeling effort was reviewed by Böning and Semtner, 2001).

A few of the major strategic debates stand out. One was the conflict between those who believed that the major issues of physical oceanography and climate lay with the inability to parametrize processes in the models, and those advocating a global quantitative description of the circulation. Thus, a strong community wished to deploy the majority of WOCE observational resources into a single ocean basin (there were advocates for the North Atlantic and the North Pacific). WOCE did endorse and carry out a number of regional process-oriented experiments, most notably the so-called Subduction Experiment in the eastern North Atlantic, and the Brazil Basin Experiment in the South Atlantic, but some of the fiercer advocates of what was sometimes called "model testing" declined further participation in the program. Another complication was the organizational separation of the Subduction Experiment from WOCE because the U.S. Office of Naval Research was interested in funding it, but did not want to be attached in any way to a program that was publicly directed at understanding climate.

Overall, the WOCE organizers generally succeeded in maintaining the global-scale, deep-water, measurement focus which had underlain the initial proposals for the program. Other specific regions had powerful proponents (high-latitude marginal seas, the Mediterranean, and so on) who simply could not be accommodated with the resources (human and observational) that were likely to be available. With hindsight, it is clear that the global ocean is so complex, with so many different dynamical regimes, and time and space scales, that few individuals are comfortable with discussions of the system as a whole. Most scientists focus their attention on particular processes, or ocean basins, and the global scale tends to be an orphan. That WOCE did not break up into a series of regional programs was one of the great accomplishments of the various steering committees. (Some of the ongoing travails of the successor CLIVAR can be understood in this context.)

Getting people to think about the global problem was not so easy, if only because the costs seemed prohibitive. Figures 12.3 and 12.4 were drawn by me in early 1982 with a ruler and marking pen, simply to permit a rough calculation of what a global hydrographic program would cost. The reaction that "we could never afford that" was addressed by dividing the number of sections by about 5 years, and by the number of institutions around the world capable of doing high-quality hydrographic work. Although not cheap or easy, it was eventually agreed that such a program was indeed manageable. The final WOCE hydrographic coverage is qualitatively somewhat like what was sketched. (At least one hydrographer had difficulty distinguishing a scale analysis for cost purposes from a detailed plan and was so affronted by it, he assured me that he was going to make certain that *none* of these lines would be measured!)

The balancing of costs against scientific benefit, absent any quantitative tools for determining the latter, was a major difficulty. Was it important, and worth the financial costs and human effort, to deploy current meter moorings in the central South Pacific Ocean where such measurements had never been made? Even today, with far more capable models and ability to determine data impact on various estimated quantities, such questions are rarely posed and answered quantitatively. Inevitably, WOCE *in situ* observations were determined through complex negotiations in national and international meetings that gave great weight to the presence of people who had particular observational capabilities, who wished to participate, and were capable of bringing national resources with them to the program. (Funds under the control of the international WOCE steering groups were limited to less than what was necessary to maintain a coordinating office in Wormley, U.K., and travel for the steering group members.) A prime example of the debates taking place concerned the high costs of adding a major transient and "exotic" tracer program to the WOCE hydrographic survey. B. Warren (WHOI) had written a letter, 9 February 1987, to the U.S. cochairmen (W. Nowlin, C. Wunsch) questioning whether the scientific payback from such measurements could justify the very considerable expense, and whose most immediate impact would be to reduce the spatial coverage of the program. Fierce debate ensued between proponents and skeptics of such measurements. Although some of the more burdensome of the proposed measurements were dropped (argon-39 measurements, notably, would have required huge sample volumes—several tons each—and the water could only be analyzed in Bern, Switzerland), a largely political decision was made that without tracer community participation and enthusiasm, the hydrographic program was unlikely to be fundable. A major tracer program thus was carried out. (It would now be possible to answer the question of whether the scientific return from the tracer measurements was worth the cost and overall spatial and temporal coverage reduction, but to my knowledge, no such study has been done. Sleeping dogs are probably best left alone.)

Getting satellites flown (the WOCE planners sought not only what became TOPEX/POSEIDON; but also the ERS-1 satellite; a scatterometer to measure the windfield; as well as a gravity mission to provide for absolute altimetry) proved

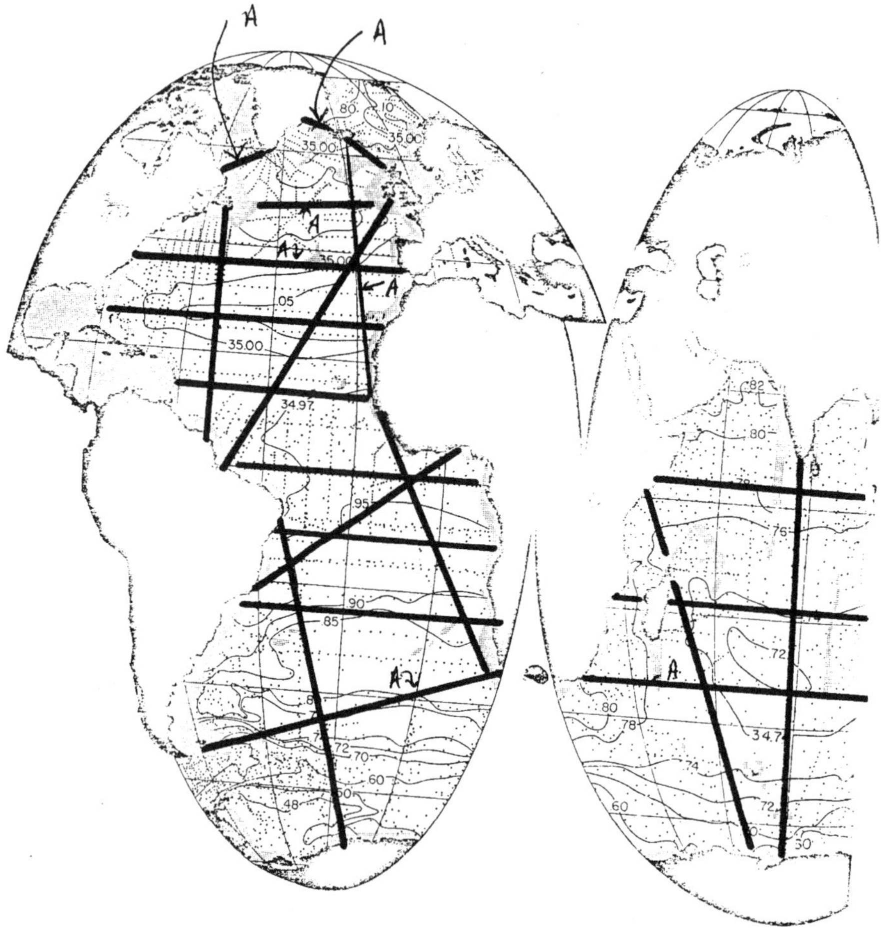

Figures 12.3 and 12.4. Sketches (Wunsch, 1984) of a global hydrographic program. Lines drawn with a straight edge on top of Reid's (1981) salinity field. The figure was intended only for estimation of costs, but was taken by some to represent the actual plan—which took many years to define in detail.

to be a very complicated story in its own right. National space agencies, such as the U.S. NASA and the French CNES, have their own politics, dynamics, and a multiplicity of constituencies. International space agencies (ESA) are immensely complicated organizations attempting to respond to diverse national pressures and priorities. Advocates of WOCE and the various satellite missions undertook a long negotiation process to attempt to provide a simultaneous global *in situ* field program along with concurrent flight of the requisite spacecraft. Although I will not attempt to describe the details of this process here, much of the strategy consisted of telling oceanographic funding agencies that WOCE had to be done within a finite time

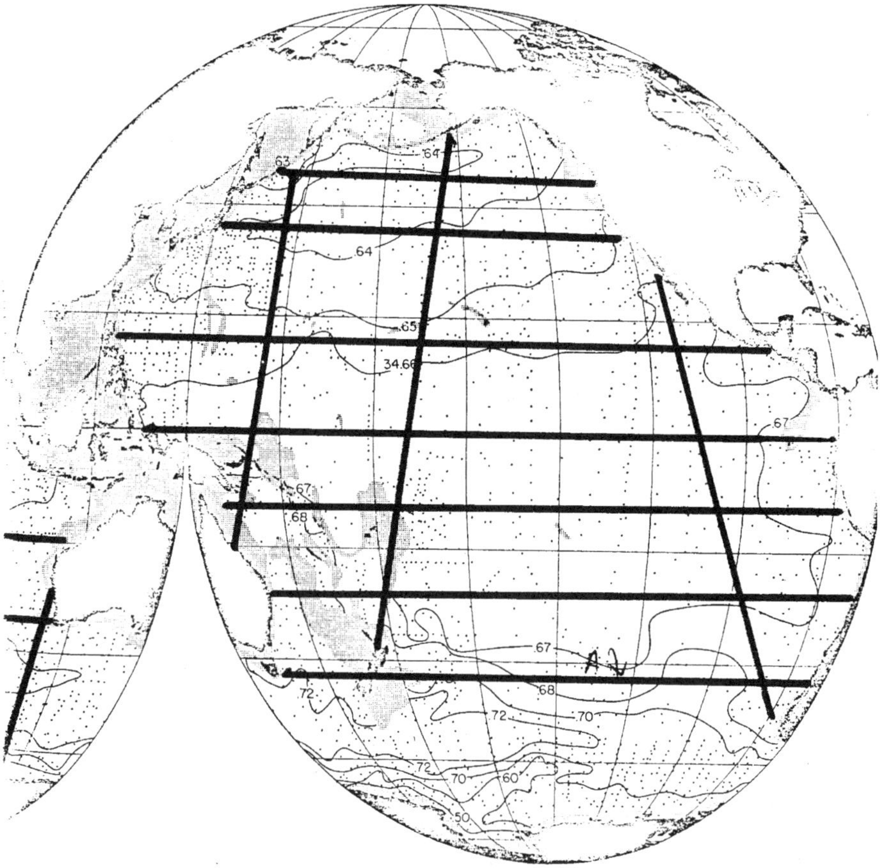

interval so as to take advantage of the independently funded satellite missions, and simultaneously telling the space agencies that the satellites had to be flown in a finite time window to take advantage of the independently funded *in situ*, WOCE program. The strategy worked for altimetry; only marginally for scatterometery; and failed for gravity missions which are only now becoming reality. (Cost estimates for WOCE vary greatly depending upon whether one includes the satellite expenditures. During the planning process, some oceanographers never did seem to understand that if an oceanographic satellite such as TOPEX/POSEIDON were cancelled, the resulting funds would *not* be available for *in situ* observations. Considerable acrimony existed over this point. The effort to fly a high-precision altimetric satellite was extremely unpopular with much of the physical oceanographic community, many of whom regarded it as a colossal waste of money. This widespread skepticism was artfully concealed, in particular, from NASA management.)

SOME AFTERTHOUGHTS

WOCE was a watershed in the history of oceanography, and it is difficult to envision any similar program being carried out ever again: with WOCE, the era of pure exploration of the fluid ocean largely ended. One could no longer point (as we did in our planning documents) to large regions of frequency/wave number space where there was no information at all (e.g., "how much does the ocean vary on time scales of 3 months on spatial scales of 2000 km?" was an unaddressable question. Now we can give very precise answers for much of the system.). We are now in an era where spatial scales ranging from millimeters to 10,000 km, and global-scale temporal variations of days to decades, have been measured. Not all such scales have been measured in all geographical regions, but there is no longer a "mare incognita" of the same extent. Figure 12.5 shows the completely schematic frequency wave number diagram, used by the TOPEX Science Working Group (1981) to discuss the problems of sampling the ocean. Units were carefully omitted from the contours because it was not possible to make a quantitative estimate of the spectrum at that time. The report argued that apart from limited knowledge of the mesoscale in the North Atlantic, and some

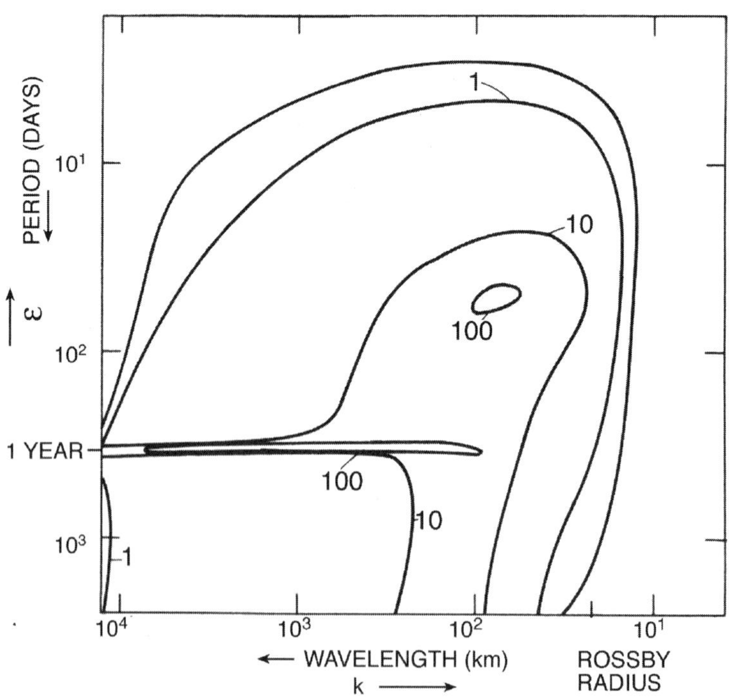

Figure 12.5. Schematic frequency wave number diagram, without units, constructed by the TOPEX Science Working Group (1981).

knowledge of the annual cycle of sea level from tide gauges (Patullo *et al.*, 1955), almost none of the spectrum had ever been measured.

The very success of WOCE has led to present difficulties in further pursuing classical physical oceanography. Major issues now lie with determining how to maintain global-scale measurements for indefinite periods—largely taking them out of the realm of possibility for academic oceanographers working on three- to-five-year grant and six-year tenure cycles. Although many processes are still poorly understood, we now have models on both regional and global scales that when constrained to our WOCE-generated data sets clearly have skill, and are useful in a way that was not true 20 years ago (e.g., Stammer *et al.*, 2002). The increasing regional focus of much of the literature is a paradoxical outcome of the success of the global experiment—much interest now lies with specific regional variations in physical processes (e.g., tidal mixing variations) relative to the presumptive global averages.[8]

Before WOCE, one could, for example, obtain funding to study the monsoon regime of the western Indian Ocean for a year or two. What is now known of that region, from WOCE and parallel efforts, leads to the conclusions that many years, and probably many decades, of observation will be required to make a qualitative improvement in existing understanding—because of the very strong interannual variability that must be accounted for.

Much of what we now take for granted (e.g., global altimetric maps of variability every few days) was science fiction 20 years ago. Students entering the field since about 1995 can, and should, take for granted the existence of a global data base, ongoing efforts to estimate the time-evolving ocean with realistic-seeming models, and a wide variety of remarkable instruments that emerged from WOCE (or during the period in which WOCE evolved). But thousands of people from dozens of countries made it all possible, and sometimes it is worth looking back to appreciate that we do make some forward progress.

What of CAGE? It was a good idea, and to a great extent, WOCE subsumed it.[9] As a token of how far we have progressed, Figure 12.6 shows the global transport

[8] I am aware that these are sweeping generalizations to which there are many caveats and exceptions, but it is also true that there has been a qualitative change in the way we do large-scale physical oceanography.

[9] Bob Stewart was not particularly unhappy that his CAGE proposal was not *per se*, carried out. He was a powerful supporter of WOCE, and efforts such as his were extremely important in gaining acceptance for the program. What I did not realize at the time was that Stewart and other prominent physical oceanographers were pleased with the WOCE proposal because it allowed them to shove aside a persistent Soviet Union "Sections" proposal. A very senior Russian meteorologist, G. Marchuk, had for years been advocating at international meetings a program for committing all oceanographic ships to repeated hydrographic sections in regions that Marchuk claimed to have identified as controlling weather. Stewart, H. Stommel, and others were fearful that the plan was going to gain acceptance and absorb much of the world's oceanographic efforts—all based upon one powerful man's insistence. "Sections" was again proposed at the same meeting where CAGE and WOCE were originally discussed, but was brushed aside. Whatever misgivings Stewart *et al.* may have had about WOCE (and Stommel surely did), there was some hope that something useful would come of it. Henry Stommel, who had been my thesis adviser and remained a good friend, privately strongly deprecated the idea of WOCE, resorting on more than one

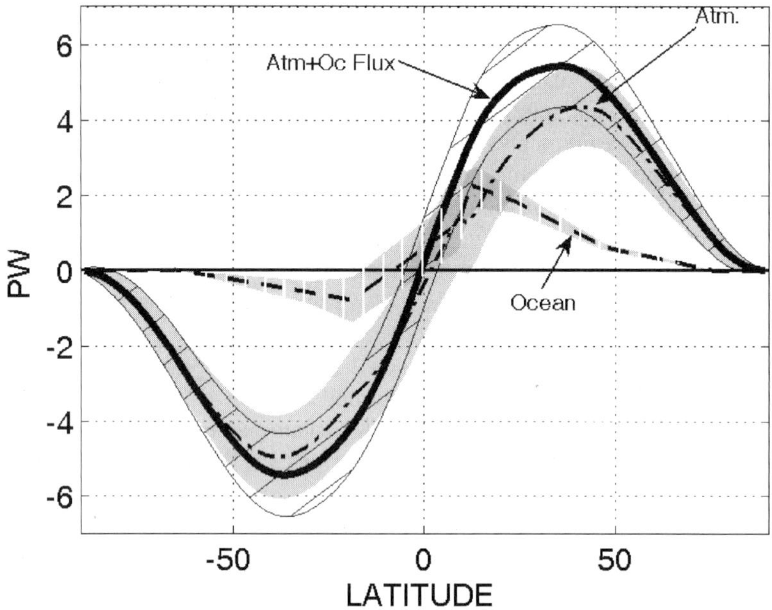

Figure 12.6. Solid curve is estimated transport of the combined ocean and atmosphere as calculated from the net outgoing radiation as measured by the ERBE satellites. Dashed line is the estimate from WOCE hydrography (primarily Ganachaud and Wunsch, 2002, but supplemented by other estimates) of the meridional flux of heat by the ocean. Dash-dot line is the inferred atmospheric transport as a residual of the total and ocean. One standard deviation error bars are shown. (From Wunsch, 2005.) Without WOCE, such calculations would not have been possible for many years, perhaps never.

of heat by the ocean and atmosphere. The ocean component was computed from the WOCE hydrographic long lines; the atmospheric component was estimated as the residual left when the oceanic component is subtracted from the net outgoing earth radiation. Twenty years ago, computing the atmosphere as a residual of measurements of the ocean would have been a laughable goal—indeed, the best oceanographic estimates were done the opposite way—with the ocean calculated as a residual of the atmosphere. Whatever the errors remaining in Figure 12.6 (and they are significant), WOCE made physical oceanography and climate a mature, quantitative subject quite unlike what it was in 1980. The challenge now is to sustain the observations and

occasion to asking my wife why I was trying to destroy my career? Sadly, he died just as the program got underway. I like to think that in the end he would have been pleased by how much we have learned about the ocean. Toward the end of his life, he did offer a kind of apology—saying that he thought WOCE was inevitable—in the same way that MODE had been an inevitable program. This comment can be interpreted in several ways!

Bob Stewart was for many years deeply worried about the Soviet initiative, to the point that he published (Stewart and Braarud, 1969) an essay explaining why the effort did not make sense. The Soviet push continued, however (letter from R. W. Stewart to C. Wunsch, 4 April 1983).

model/data synthesis efforts so that our successors will not be as blind as we were in 1980 to the time-evolving ocean.

A number of elements of WOCE failed to come to pass. As already noted, the scatterometer-wind satellite did not fly until the program was almost over (and then failed prematurely), and no gravity mission appeared until the launch of GRACE in March 2002. Efforts to define a full-water column equatorial ocean observation component came to little with the focus of the tropical oceanographic community on the upper ocean alone (to this day, there are no instruments on the TOGA-TAO array—its observational legacy—below 500 m). Some proposed elements, e.g., the open ocean current meter moorings, were never deployed. Few oceanographers have retained an interest in studying the ocean as a whole—rather there has been a reversion toward regional programs and processes (cf. CLIVAR). Whether the wider community will find a way to sustain the global observation network (now primarily satellites, the ARGO float program, the diminishing XBT coverage, and intermittent revisits of WOCE hydrographic lines) is one of the major challenges for the future. Recognition that it needs to be done may perhaps be the ultimate legacy of WOCE. There is little doubt, however, that without WOCE, oceanography would be a very different subject than it is today.

It is worth remembering that WOCE was an extremely controversial program, although the disputes have largely faded from memory. Anyone motivated to organize a future observational experiment of equivalent scope may perhaps be comforted to realize that ultimate success means that the inevitable, if painful, dissent will be forgotten. In the context of the present, more complex situation, in which much more is now known about the ocean and new international bureaucracies exist, an important lesson is that WOCE was a "bottom-up" program—a critical mass of individual working scientists sought to create the program because they believed it scientifically necessary, and because they personally wished to work with the resulting observations.[10] (A few forward-looking scientists recognized that the time span of the program would exceed the span of their own professional careers—they nonetheless worked for its creation because they recognized its scientific importance.) WOCE was born at a time when it was scientifically ripe. Later, "top-down" initiatives arising from the national and international committee structures are often comparatively sterile in outcome because the underlying scientific motivation is secondary to programmatic structures.

ACKNOWLEDGEMENTS

WOCE was the result of efforts by thousands of people in dozens of countries around the world, including program managers, principal investigators, engineers,

[10] That WOCE arose out of the initiative of a few individual scientists eventually became difficult to perceive. Scientists coming to the program after the start of the planning process encountered a WOCE managerial bureaucracy that had been created to implement it, not create it.

technicians, secretaries, ships crews, and many others. In the spirit of a purely personal essay, I would like to particularly acknowledge the work in the earliest days of Professor Worth Nowlin (Texas A&M) without whom the U.S. contribution to WOCE would clearly have come to nought, Professor John Woods (now Imperial College) whose organizational skills were critical in the early years, and Michel Lefebvre for bringing his infectious enthusiasm and the French POSEIDON project to WOCE. Many other far-sighted individuals deserve thanks for their sometimes heroic efforts, but because I am sure to forgot someone, it seems best to simply acknowledge that the community owes a large debt to many people—who at least know who they are. Preparation of this essay was supported in part by the National Ocean Partnership (NOPP) ECCO Consortium funding, an extension of WOCE. I had helpful comments from M. Jochum, D. J. Baker, and W. Munk.

REFERENCES

Böning, C. W., and A. J. Semtner, 2001. High resolution modeling of the thermohaline and wind-driven circulation. In: *Ocean Circulation and Climate: Observing and Modeling the Global Ocean*, G. Siedler, J. Church, and J. Gould, eds., Academic Press, San Diego, 715 pp.

Cheney, R. E., 1982. Comparison data for SEASAT altimetry in the western North Atlantic. *J. Geophys. Res.* **87**, 3247–3253.

Collins, C. A., and R. H. Heinmiller, 1989. The Polymode Program. *Ocean Dev. Int. Law* **20**, 391–408.

Committee on Metrics for Global Change Research, 2005. Thinking Strategically: The Appropriate Use of Metrics for the Climate Change Science Program. Climate Research Committee, Board on Atmospheric Sciences and Climate, Division on Earth and Life Sciences, National Research Council, National Academies Press, Washington.

Davis, R. E., D. C. Webb, L. A. Regier, and J. Dufour, 1992. The autonomous Lagrangian circulation explorer (ALACE). *J. Atmos. Oceanic Technol.* **9**, 264–285.

Dobson, F. W., F. P. Bretherton, D. M. Burridge, J. Crease, and E. B. Krauss, 1982. Report of the JSC/CCCO CAGE Experiment. A Feasibility Study. World Climate Programme Report, WCP-22, Geneva, 95 pp.

Fuglister, F. C., 1960. Atlantic Ocean Atlas of Temperature and Salinity Profiles and Data from the International Geophysical Year of 1957–1958. Woods Hole Oceanographic Institution Atlas Series: I, 209 pp.

Ganachaud, A., and C. Wunsch, 2002. Large-scale ocean heat and freshwater transports during the World Ocean Circulation Experiment. *J. Climate*, **16**, 696–705.

Garrett, C., and C. Wunsch, 1984. A Celebration in Geophysics and Oceanography—1982. In Honor of Walter Munk on His 65th Birthday October 19, 1982. Scripps Institution of Oceanography Reference Series 84–5, March 1984, La Jolla.

Halpern, D., 1996. Visiting TOGA's past. *Bull. Am. Meteorol. Soc.* **77**, 233–242.

Holland, W. R., and L. B. Lin, 1975. On the generation of mesoscale eddies and their contribution to the general circulation. I. A preliminary numerical experiment. *J. Phys. Oceanogr.* **5**, 642–657.

Jeffreys, H., 1933. The function of cyclones in the general circulation. Proces Verbaux de l'Assoc. de Meteorologic. UGGI, Lisbon, Part 2, 219–233, also in Collected Papers, V, 257–269.

Kerr, R. A., 1991. Greenhouse bandwagon rolls on. *Science* 23 August, 845. See Letters to the Editor, by C. Wunsch and by J. McCarthy. *Science* 18 October 1991, 357.

Lorenz, E. N., 1967. The Nature and Theory of the General Circulation of the Atmosphere. World Meteorological. Organization, Geneva, WMO No. 218, T. P. 115, 161 pp.

MODE Group, The, 1978. The Mid-Ocean Dynamics Experiment. *Deep-Sea Res.* **25**, 859–910.

Monin, A. S., V. M. Kamenkovich, and V. G. Kort, 1977. *Variability of the Oceans*. Wiley, New York, 241 pp.

Moore, G. E., 1965. Moore's law. *Electronics* **38**, 114.

Munk, W., and C. Wunsch, 1979. Ocean acoustic tomography: A scheme for large scale monitoring. *Deep-Sea Res.* **26A**, 439–464.

Munk, W., and C. Wunsch, 1982. Observing the ocean in the 1990s. *Philos. Trans. R. Soc. London Ser. A* **307**, 439–464.

Patullo, J. G., W. H. Munk, R. Revelle, and E. Strong, 1955. The seasonal oscillation in sea level. *J. Mar. Res.* **14**, 88–155.

Reid, J. L., 1981. On the mid-depth circulation of the world ocean. In: *Evolution of Physical Oceanography. Scientific Surveys in Honor of Henry Stommel*, B. A. Warren and C. Wunsch, eds., MIT Press, Cambridge, MA, 70–111.

Roemmich, D., and C. Wunsch, 1985. Two transatlantic sections: Meridional circulation and heat flux in the subtropical North Atlantic Ocean. *Deep-Sea Res.* **32**, 619–664.

Rossby, T., D. Dorson, and J. Fontaine, 1986. The RAFOS system. *J. Atmos. Oceanic Technol.* **3**, 672–689.

Stammer, D., C. Wunsch, R. Giering, C. Eckert, P. Heimbach, J. Marotzke, A. Adcroft, C. Hill, and J. Marshall, 2002. The global ocean state during 1992–1997, estimated from ocean observations and a general circulation model. Part I. Methodology and estimated state. *J. Geophys. Res.*, DOI: 10.1029/2001JC000888.

Stewart, R. W., and T. Braarud, 1969. Editorial introduction. *Prog. Oceanogr.* **5**, vii–xi.

Thompson, B. J., J. Crease, and J. Gould, 2001. The origins, development and conduct of WOCE. In *Ocean Circulation and Climate: Observing and Modeling the Global Ocean*, G. Siedler, J. Church, and J. Gould, eds., Academic Press, San Diego, 31–43.

TOPEX Science Working Group, 1981. Satellite Altimetric Measurements of the Ocean. National Aeronautics and Space Administration, Jet Propulsion Laboratory, California Institute of Technology, Pasadena, 78 pp.

Wunsch, C., 1984. A water mass conversion experiment for WOCE. Unpublished study document, 9 February 1984, 10 pp.

Wunsch, C., 1996. *The Ocean Circulation Inverse Problem*. Cambridge University Press, Cambridge, 437 pp.

Wunsch, C., 2005. The total meridional heat flux and its oceanic and atmospheric partition. *J. Climate*, **18**, 4374–4380.

Wunsch, C., and E. M. Gaposchkin, 1980. On using satellite altimetry to determine the general circulation of the oceans with application to geoid improvement. *Rev. Geophys. Space Phys.* **18**, 725–745.

Chapter 13

Interview with Klaus Wyrtki, 25 February 1999

HANS VON STORCH, JÜRGEN SÜNDERMANN, AND LORENZ MAGAARD

I

Dr. Wyrtki, you started your studies in Marburg just after the war and then you continued in Kiel. Could you explain and tell us a little bit about your university studies?

It was after the war in 1945 and I traveled up and down through Western Germany to find admission at a university. I finally succeeded in Marburg. When I was asked what to study I chose physics and mathematics because ship building what I intended to study was no longer being taught in Germany. After a while I got interested in applications and I read books about meteorology and in doing so I found out that oceanography existed. I read Defant's "Dynamische Ozeanographie" and other books. Eventually I went to my geography professor—I think his name was Schmitthenner—and asked him where oceanography was taught. He said that there was a famous institute in Berlin, but that it was bombed out and that most of the people had probably moved to Kiel. In the summer of 1947 I went up to Kiel to visit the Institut für Meereskunde[1]. When I climbed up to the tower of the villa, Hohenbergstraße 2,

[1] Institute of Oceanography.

where the Institut für Meereskunde as well as the Geological Institute were located, I found Georg Wüst and I told him my story. When I had finished he said, "well that's nice. Now I have a student". That's how it started with me. He arranged for an exchange of student places which was possible at that time. In the summer of 1948 I went up to Kiel.

There comes to mind the story about my dissertation. After a year or so I asked Wüst, I would like to make a Ph.D. and he said, "fine, let us do. There is someone in the German Hydrographic Institute who has an instrument that measures turbidity in the ocean and you just take the instrument and go out to sea and measure more often than anybody has measured with it. And you will find something new.[2] Dr. Krey has worked with the instrument, go and see him." I had to calibrate the instrument. When talking with Krey about it, he gave me two big volumes of colloid chemistry which I had never heard anything about. I put them in the lowest drawer on my desk and never opened them until I had my Ph.D.[3] I didn't intend to do anything about chemistry, but he thought that the substances that were in the ocean and would be measured by the light were mainly of chemical nature.

Anyway let us go on. You asked what I learned from Wüst. It's basically the general overview, to look at large connections, not at the details, but to integrate things, to see the big picture.

You asked for the little story about an attachment to a bicycle. We students were somewhat annoyed that we had to carry boxes of water samples and instruments from the institute to the research ship and back. We wanted some easier way of transportation. Wüst approved of that and told us to buy a little cart to hang behind a bicycle. The university administration did not approve that. It was not a scientific instrument. We came to use the name "transporteur" which is actually a measuring device used by surveyors to fix angles on charts. We submitted that to the administration; it was approved as 'transporteur' and the bicycle dealer actually sold us one of the two wheel carts to hang behind a bicycle. That is the way, how we misled the administration.

Thank you very much for this advice. We keep that in mind.

You keep that in mind. That is good.

You finished your studies at the university with receiving your Ph.D. Does it mean that you never had a classical examination at the university?

Not really, except for a few little examinations. As a student in the natural sciences I had to take one course in Germanistics. It was a seminar on an obscure German poet, who had written a lot of novels and we were supposed to read all these novels. When examination came I had read none, not a single one. About twelve students were sitting around a big table with the professor and he started to ask the first one

---

[2] See also page 212.

[3] Wyrtki, K., 1950: Über die Verteilung der Trübung in den Wassermassen der Beltsee und ihren Zusammenhang mit den hydrographischen Faktoren. Ph.D. dissertation, Univ. Kiel, FRG, 49 pp.

about one novel, the second one about the second novel. I saw that it wouldn't go very smoothly, and I was sitting in the middle. When he was at the fifth, I interrupted him. I thought, attack is the best defense, and discussed with him something about the ethics of the knights, die Ethik der Ritter, because one of the novels was about the knights. We discussed that for a while, then he took the next student, then he skipped me and he went on and when we finally got our slips, it said 'good', that was fine, that was my examination. This was a little footnote of my student days. There was of course a final examination for my Ph.D.

Your university studies were significantly different from today. Today everything is regulated, more or less. Do you find that your way of taking the university was somewhat better?

It was a wonderful freedom that we had. You could study, you could not study. You could do what you wanted. You had to have responsibility. That wasn't taken away from you. If you failed, you failed. You were out. Today we are giving remedial courses. Students shouldn't get remedial courses, they should be thrown out. That's my opinion. That's not the university opinion.

After I had my Ph.D. I had a very short stint in Hamburg. At that time Dietrich had a position with the British Navy to oversee German oceanography and to collect material from the war and to hand it over to the British. Dietrich got a university appointment at that time. There were six months of salary left in that position which was under the control of a British admiral Carruthers. I moved to Hamburg for six months and my room was one floor above Bönecke, the director, because I was the representative of His Majesty, the Queen. From time to time Bönecke gave me a call, "Wyrtki, kommen Sie runter,[4] you have to sign a document on behalf of His Majesty the Queen". He was smiling about these things. That is the way things go.

You were asking about salaries. When I was research assistant, I had 300 marks. That was barely sufficient to get along as a student, and suddenly with my appointment in Hamburg, I got 800 marks and I felt like a king. I suddenly had everything I wanted.

What did you do with all the money?

Amazing. At that time you still had to buy clothing, you could go out a little bit. You could live.

We should compare that with how much you had to pay for a car, for a Volkswagen, for instance.

A car at that time, about 1500 marks, Volkswagen Beetle. It's amazing, but that's it.

After the six months in Hamburg I returned to Kiel and I got a Forschungsauftrag von der Notgemeinschaft Deutscher Wissenschaften.[5] That was for the studies of the

[4] Here, Wyrtki changed spontaneously into German: "come down".

[5] A research grant from the German Science Foundation.

water exchange between the Baltic and the North Sea which I did then for three years. We made a lot of measurements in the Fehmarn Belt and elsewhere, with paddle wheel current meters to study water movements. I analyzed data. Interpretation of data was always what interested me.

When the three years of the research grant were finished I was looking for a job. Neither Wüst nor Bönecke had one for me. A friend of mine, Willi Brogmus, got a letter from Indonesia asking whether he wanted to come to Indonesia as a scientist.

May I ask something between before you go to Indonesia? I noticed that you had this project from German Science Foundation. Who were the reviewers in those days? There were only very few oceanographers in Germany.

Honestly speaking, I think it was done on the recommendation of Wüst. He was the professor at that time. There were no reviewers. Reviewing was an unknown matter, also reviewing for journals was not in existence. When a senior professor told a journal to publish something it was published.

Was that only so in Germany?

I would say that was in France and elsewhere, maybe not in England.

Could you say a few names? What persons worked in oceanography just after the war at that time?

After the war there was Hansen, at the DHI,[6] Joseph in physical oceanography, there was of course Dietrich. There was Neumann and Roll at the Institute of Geophysics at Hamburg. There were some more people. Tomczak, the father. Weidemann was assistant to Wüst.

They mainly worked in the German Hydrographic Institute?

Yes.

We stopped at Willi Brogmus. He declared he would rather go to the North Pole than into the tropics. So he gave me that letter. I wrote to Indonesia, a few months later I was on the way to Indonesia. This went all pretty easy. When I arrived in Indonesia, they were phasing out the Dutch at that time and they were looking for other people. Since Germany had no colonial attachments we were somewhat welcome in these countries. In Indonesia I found myself not only the only scientist in the institute, because all the Dutch had left, but I was also the director of it. I had a research vessel of about 200 tons, a nice yacht type vessel, the "Samudera". I made many voyages with it, with very little instrumentation. We did a few surveys with Nansen bottles down to a few hundred meters but could not reach the deep sea basins in Indonesia because of a lack of a long wire, and that restricted us to the surface layers.

---

[6] Deutsches Hydrographisches Institut = German Hydrographic Institute in Hamburg.

On board "Samudera", 1955

I discovered there was a lot of actual information about these waters that had never been summarized. I started to work on a book, the physical oceanography of the Southeast Asian waters; it became known as the NAGA Report[7] later on when it was published at Scripps. I wrote that book on many long voyages through the Indonesian waters. That proved actually quite a hit, because the information about these waters had never been summarized and it remained a valuable reference for decades because the Indonesians were very hesitant in the decades that followed to let foreigners doing research in their waters. We come back to that when we talk about international cooperation.[8]

Did you find at that time the Indonesian through-flow?

Yes, when analyzing the data from both the Dana and the Snellins expeditions. The Snellins expedition was not completely published by that time. I could analyze existing sea level data, I could make dynamic calculation, both in the Pacific and in the Indian Ocean. I could identify the fact that there was a pressure difference between the two. I analyzed surface circulation which indicated that there was a monsoon dependent through-flow. That was the start of that type of research.

After your time in Indonesia you went to Australia.

From Indonesia I was sent to Tokyo, in 1955 for a UNESCO conference. There were all the famous oceanographers, including Roger Revelle, Deacon from England, Hidaka, Bönecke and so on. That time I met Roger Revelle and that turned out to be a very profitable meeting in the long run. We talked quite a while and I met Roger Revelle

---

7 Wyrtki, K., 1961: Physical oceanography of the southeast Asian waters. Univ. Calif., NAGA Rept., No. 2, 195 pp.

8 See page 221.

again at the Pacific Science Congress in Bangkok in 1957 when I was on the way back to Germany from Indonesia.

I actually gave up my position in Indonesia, and didn't extend my three years contract because there started a civil war in Sumatra at that time and conditions were restless. I had several months of vacation coming up anyway and a free trip back to Germany. I went via Bangkok, where I met Roger Revelle again, I met Townsend Cromwell, the discoverer of the equatorial undercurrent, and other people.

When I came back to Germany in 1958, Bönecke had lined up a job for me. That was in Monaco. Bönecke at that time was promoting the general bathymetric charts of the world. The International Hydrographic Bureau in Monaco was supposed to do them. I went down to Monaco for about 6 months. This was basically a post office. It was scientifically not challenging in any way and for that reason I didn't stay there. I could have stayed, but it was a dead end career. Recognizing that early enough I looked into other positions available.

There was one position in Australia offered in 'Nature'. I applied for it and actually got the position. After the Monaco stay was over, I went in November 1958 to Australia. There in Australia I had a wonderful time with the CSIRO Division of Fisheries and Oceanography. It was similar to what in Germany are the Max-Planck Institutions. That means research institutions granted by the government. I had very nice colleagues. We had Neil Brown who with Bruce Hamon constructed the first CTD and we tried it out at sea. We had David Rocheford. There was the International Indian Ocean Expedition going on in which I did not participate because my work was on the oceanography in the Tasman and Coral Sea. My interest developed at that time into Antarctic circulation. That was really following in the footsteps of Wüst, deep ocean circulation and the Antarctic water ring that connects the deep circulation of all the oceans.

Did you know that at that time already?

This was known by Sverdrup and by Deacon. Science is always a progress. You want to know something better. In fact many good ideas you get just from reading older papers. What kind of speculations good scientists make about the things that are unknown. That are not readily accessible to them. The data are limiting. If you look up their ideas and follow them through with new data you are probably onto something. That is when I wrote the papers on thermohaline circulation and on the oxygen minima in the oceans.[9] The oxygen minimum paper has been widely used by geochemists to explain the distribution of properties.

That was the time when it became clear to me that vertical movements are the main links in ocean circulation—like the Antarctic upwelling, like the vertical movements in the deep ocean basins that must bring slowly up water to the surface

---

[9] Wyrtki, K., 1961: The thermohaline circulation in relation to general circulation in the oceans. *Deep-Sea Res.*, **8** (1), 39–64.
Wyrtki, K., 1962: The oxygen minima in relation to ocean circulation. *Deep-Sea Res.*, **9**, 11–23.

and are counteracted by vertical diffusion. All these problems were at that time addressed.[10]

At the same time it became quite clear that surface circulation in contrast to deep circulation was very variable, as we could see from surveys that we made in the East Australia Current.

While I was in Australia a colleague of mine, a zoologist, spent a sabbatical at Scripps. When he came back he said, "Klaus, the people at Scripps want your curriculum vitae". I sent them my curriculum vitae. Of course in the curriculum vitae you had to give references. One of the references was Georg Wüst, who at that time was at Columbia University. After about two weeks I got a job offer from Columbia University. That went that fast.

I tried to find out what the future would offer. At Scripps I would belong to a tuna research program that stretched all the way from California to Peru, throughout the eastern tropical Pacific investigating the environment of the tuna population. At Columbia I would be assigned to a new research ship, the ELTANIN, and I would go into the Antarctic Ocean. Arnold Gordon eventually got the job, because I said, "no, no. No Antarctic Ocean, no seasickness, no roaring forties, I stay in the tropics". After Indonesia I was spoiled, I didn't want to go back to the cold climate, so Scripps institution won.

Likely Wüst was disappointed.

Wüst was disappointed, of course, but he got Arnold Gordon. That was fine.

On the way from Australia to California I stopped in Hawaii for a Pacific Science Congress. That was the Pacific Science Congress during which the cornerstone for the Hawaii Institute of Geophysics was being laid but at that time I was not aware that I would finish up there.

So, I came to Scripps and the work there was most interesting. It was not data taking, other people were doing that. It was studying the upper ocean variability. At that time it had become clear that fisheries and long-term weather prediction are dependent on oceanographic knowledge on a real-time basis. One needed to know what happened in the ocean from month to month and from year to year in order to explain how the environment reacts.

Did you learn also something from biology at that time? Or from biologists?

I didn't have to know much, I had enough fishery biologists around me and we had very close interaction with the people who were doing the tuna research in biology, the tuna marketing and catching, the fishery people actually running the fishing fleets. We gave them BTs—that was the study on the Costa Rica Dome,[11] on upwelling, where cold water comes up to within 10 meters and where the tuna boats can put the

---

[10] Wyrtki, K., 1963. The horizontal and vertical field of motion in the Peru Current. *Bull. Scripps Inst. Oceanogr.* Univ. Calif., **8** (4), 313–346.

[11] Wyrtki, K., 1964: Upwelling in the Costa Rica Dome. *Fish. Bull.*, **63** (2), 355–372.

big nets around a whole school of tunas and fishes, and get tens of tons of tuna out. The Peruvian fishery was growing at that time, at a tremendous rate.

It was a very exciting and productive era, I met Jakob Bjerknes at that time, he came often down from Los Angeles. My neighbors were Jonny Knauss, Joe Reid, Wooster, all these people, we were all together there; Benny Schäfer was the director of fisheries research.

Was that the time when you started using a computer?

Yes, that was the time when we first wanted to get maps of surface temperature on a monthly basis and if you do that, you need data in a short time. Ship observations were collected. They came in by radio through the meteorological network and you had to collect and to process them. We had the task with thousands of observations that we wanted to map and so one day I said we have to use computers and we looked for someone who could do computer programming. We found a graduate chemistry student. He came up to me and I explained to him what we needed, he said that he could do that, but I would have to write him some instructions. In a couple of days I wrote down the instructions, and when he came back the next time, I handed him the sheet and he looked at the sheet, then he looked at me and he said, "oh, you have written a computer program". This was a list of instructions on how to go in sequence through the mass of data. I had no idea about computer programming at that time.

Did you yourself any programming?

No, never.

At times I had up to four, five computer programmers working for. I knew what goes in and what comes out, but that was it. Like with an appendix. I don't start to study medicine when I want my appendix out.

Did you begin to use a personal computer for writing and email?

Yes, in the NORPAX project we were among the first to use email, because we were on the Office of Naval Research circuit. For the Test Shuttle we used it as early as 1975. That was "telemail". My secretary used it every morning.

But you did not use it yourself, you did not type yourself?

No.

Another thing. My first computer programmer was hired for the Indian Ocean Atlas, it was done largely by computer. Then she became pregnant and she retired for a year and then she wanted her job back and I took her back with great welcome. Then she got her second baby and she wanted to work at home and we bought her a little computer, with which she could use her home telephone and connect to the university computer. So, she could work at home while waiting for the baby. These were the first explorations in computer. It was an exciting time.

Now Scripps. Why I got out of Scripps? The answer to that is very simple. In Scripps at that time—it has changed by now—there were two sorts of people,

Jan.16, 1972 Honolulu Star Bulletin

Wyrtki, atlas and computerized plotter.

researchers and professors. When you were a researcher, you never could become a professor.

You did not know this before?

I had no idea of the structure of an American institution. But this was general —that was the case in Woods Hole, that was the case at Columbia, Lamont, New York University, Miami. In most of the institutions, this was the situation. Since my goal was really to become a professor, to teach, to do research, I was very happy, when one morning someone knocked at my door in La Jolla and introduced himself as being the acting chairman of the new oceanography department in Hawaii. This fellow, who became later president of Texas University, was the first department chairman; his toys were analogue computers. He knocked at my door and made me an offer and I said, "yes, I come". And so I moved to Hawaii in the summer of 1964.

In the first few years in Hawaii, George Wollard was the director of the Institute of Geophysics, and money was flowing easily—we had Office of Naval Research contracts to do current measurements, contracts to do current measurements around the islands, to study islands circulation and heat advection in the North Pacific—but I started with a project that I always wanted to do, namely, investigating the circulation of the Indian Ocean. I wrote a proposal to the National Science Foundation to make the Indian Ocean Atlas on the physical oceanography. That was basically my main activity from the time of my arrival here to 1970. It was essentially in the tradition of Wüst, studying the deep circulation. There were two motivations. The deep circulation was per se of interest, but the deep circulation was basically considered stationary: once you know it you know it for the century, at least. But at Scripps I had learned how fast the upper ocean moves and that it is necessary to study the changes that are going on within weeks and months. For that reason I concentrated the work on the Indian Ocean Atlas on the study of the annual variation, which is of course natural for the Indian Ocean because of the monsoons. But if you do these things you are getting new results.

By the way, that was something I learned from Wüst: "if you take a new instrument or measure something more frequently, you will find something new." This is a basic principle and this is how my Ph.D. thesis came into being.[12]

There was no idea what you will find?

There was no idea what one might find. You take a new instrument, measure more frequently than anybody before you and you are going to find something. This was the philosophy. For instance, if everybody looks at the mean stationary state, then you look at the variability and you will get something new. In this way I found most interesting things.

You have here in your list[13] the question 'Wie entsteht wissenschaftlicher Fortschritt?'[14] and you list four items 'Förderung', 'Gelegenheit', 'Personen', 'Zufall'[15]. In my opinion all items are important. But a basic prerequisite for scientific Fortschritt ist, daß man sich wundert.[16] Man wundert sich über etwas, was nicht leicht erklärbar ist. Ich habe mich über zwei Dinge gewundert, die schließlich beide zum El Niño geführt haben.

Das erste waren die Seiches in the Baltic. Eines schönen Tages—und da kommen wir wieder auf Wüst zurück—war in Kiel Hochwasser.[17] Das Hindenburgufer[18] war überflutet und am nächsten Morgen rief mich Wüst in sein Office und sagte, "Herr Wyrtki haben Sie sich das Hochwasser am Hindenburgufer angesehen?" Ich sagte, "ja, ja". "Ja, aber wir müssen doch wissen, warum das zustande kommt. Suchen Sie sich mal all die Daten zusammen und dann werden Sie das analysieren." That were

[12] see page 204.

[13] In the tentative list of questions prepared for the interview.

[14] How is scientific progress generated?

[15] Funding, opportunity, people, coincidence.

[16] Here, Dr. Wyrtki spontaneously changed into German: "a prerequisite for scientific progress is that one is wondering. One is wondering about something not easily explainable. I was amazed over two things that both finally led to El Niño. The first were the seiches in the Baltic. At a certain day the Hindenburgufer 18 in Kiel as flooded. On the next morning Wüst called me into his office and said, "Herr Wyrtki, have you seen the flooding of the Hindenburgufer?" I said, "yes, yes". "We must know how this happened. Collect all data, and analyze them."

[17] 3 December 1952; see Wyrtki, K., 1953: Die Dynamik der Wasserbewegungen im Fehmarnbelt I. *Kiel. Meeresforsch.*, **9** (2), 155–170.

[18] A promenade in Kiel at the banks of the Kiel Bight.

wind-induced seiches of the Baltic. There were southwest winds ahead of a cold front. Twelve hours later there were northeast winds, very strong behind the cold front, and the Baltic was excited; seiches were introduced, and the Baltic schwabberte, mit der bekannten 24h-Periode.[19] Seit diesem Tage, wo ich diese seiches in der Ostsee beobachtet und gesehen habe, habe ich mich gewundert, ob der große weite, offene Ozean nicht mehr schwabbert. Das war eine Fragestellung.[20]

The other thing was related to Peru. I made a current chart for the eastern tropical Pacific and I was amazed that certain currents start nearly out of nothing and end somewhere in a very diffuse way: the huge South Equatorial Current that transports fifty Sverdrups, starts of this little Peru current that transports 10 Sv— where is all the water coming from? And the South Equatorial Current ends near New Guinea in the Coral Sea and you cannot see how it ends, it disappears. Where does all the water go? This was the next question.

When making the Indian Ocean Atlas we drew maps for every month of the topography of the 20° isotherms, i.e., of the thermocline, in the Indian Ocean. It was obvious that in certain parts of the ocean the thermocline was seasonally going down and in other parts it was seasonally going up. So the idea came, if the thermocline goes down by 20 or 30 m, how much water does it really transport out of an area? I made the rough calculation and it showed that a substantial amount—10 to 20 Sverdrups— leaves Somalia and goes over to Sumatra. And so I was looking at current charts and there was the equatorial jet in the Indian Ocean, going from one area where the thermocline lifts up to the other site of the ocean where the thermocline goes down. That was really the next step on the road to El Niño.

Did you make your own measurements in the Indian Ocean?

No, that was the International Indian Ocean Expedition on which I did not participate, because at that time I was working in the Tasman and Coral Sea. But, David Rochefort, my colleague in Australia, was one of the main participants in the International Indian Ocean Expedition.

Those were the years from 1966 to 1970, when I was working on the Indian Ocean Atlas[21]. In 1971 I spent half a year at Kiel with Dietrich on a sabbatical and when I came back, climate research started. This was the International Decade of Ocean Exploration and the National Science Foundation started to fund big projects. There was GEOSECS, MODE, the Southern Ocean, NORPAX. In the beginning I participated in the NORPAX project. After having seen in the Indian Ocean, how important annual variability is, and having known from my tuna research years that year-to-year changes are quite important, I looked at the data from Hawaii and I found

[19] The Baltic wobbled with the known 24-hour period.

[20] Since that day, when I had observed the seiches in the Baltic, I was wondering whether something like that happens in the big ocean, and why the wide open ocean is not wobbling more. That was the question.

[21] Wyrtki, K., 1971: *Oceanographic Atlas of the International Indian Ocean Expedition.* National Science Foundation Publication, OCE/NSF 86-00-001Washington, DC, 531 pp.

out that we really didn't know how the big trade wind field varies from year to year. When I asked the meteorologists, they could not tell me. That is when we started to get the ship observations, the wind observations, and crunched 25 years of ship observations—there were 3 million observations at that time for the equatorial Pacific Ocean. We learned that the trade wind fields undergo massive changes from year to year. Analyzing these changes I found out that the biggest changes are not off Peru or somewhere near the Galapagos, but they are in the Central Pacific, real massive changes of the Southeast trade winds.

At the same time we were looking at the ideas of Bjerknes, who was working on the tropical ocean and tropical ocean– atmosphere interaction. There was Namias at Scripps who was working on the North Pacific—U.S. mainland interactions. In my personal case, came the insight that the fluctuations of the wind stress on the equator are producing El Niño. Of course we had to prove it, which brought the sea level data in, because the claim was that the thermocline in the western Pacific goes up and the thermocline in the eastern Pacific goes down. We could prove by means of sea level data that these two things really happen, because there is a direct relationship between sea level changes and thermocline changes. Putting these things together gave the El Niño theory and also the knowledge that was developed at that time about equatorial Kelvin waves. But it was basically an observational fact-finding, an analysis of observations and putting the pieces together.

The fact that sea level is a very convenient variable to monitor the ocean gave the impetus for establishing the sea level network in the Pacific. With this you could study dynamics—that was before TOPEX.

Has your work become more systematic over the years? You have told us that you have dealt with various interesting pieces in the first part and after you have started in Hawaii that you really zoomed in on one thing and became more and more systematic. Is that a fair description?

Yes and no, there is certainly a truth in that, but I don't think that it is intentional, it is simply based on the fact that your experience grows. You are exposed to more information; you learn about more processes and therefore you start to integrate your knowledge. Integrating knowledge is a very important thing.

So it is more or less normal, just a fact of getting older and more experienced.

It is a natural process.

Have you always been in a beginning of a new period, at a new investigation, of new phenomena in your different stations—first in Indonesia, later in Australia, then in Scripps, and finally in Hawaii?

Again, yes and no. You know you jump at opportunities. Recognizing the opportunities is important and may be part of learning. These were all natural developments—it had to come to that, once you study the variability you necessarily get into climate and into climate change. If you think on the large scale then that is a natural way to go. Most people actually differentiate. If you give a child a toy, the first action is to take it apart and scientists do the same. They see a problem and immediately they take the problem apart, into pieces. Very few scientists integrate, that means put things together.

Would that mean that you must be concerned in several topics? You got some idea on El Niño by studying the seiches in the Baltic. It is completely different phenomenon. The integration in this case was that you had the association that they might be relevant. This would mean, it will help if you are curious about many things in the ocean and study many different things for this integration.

Definitely.

There are other activities of which you are probably even more proud of than about your scientific papers.

The start of ocean monitoring—now everybody is monitoring the ocean, the big TOGA TAO array, that constantly gives you interesting data, there are satellites—you don't believe what fights we had to get funding for ocean monitoring. While we argued "we need to observe the same thing year after year, because only if we do that we see changes. We need to know the ocean month after month, if we want to have weather prediction. We cannot go out once every five years and make an experiment. You need to monitor," there was a constant fight about ocean monitoring. I am very proud about the fact that I was involved in that and was very vigorously participating in this fight.

Another thing is free data exchange. I don't know how often I preached when I was chairman of NORPAX "in meteorology data are instantly available. Whenever a radiosonde is launched, the next minute the data go on the radio and into the World Weather Watch." Oceanographers keep their little black boxes and the data they have in them for years in their laboratories and don't want to relinquish them. Data have to be available, in particular if you want to make forecasts.

Your life up to the Hawaii occupation was very much changing. You always changed. Why did you remain after that so long in Hawaii?

I had three years Indonesia, three years Australia, three years Scripps. People were watching, if I have three years Hawaii, too. Hawaii is too nice to leave it. It is the best place in the world to live, I enjoyed the years thoroughly—certainly I have no desire to change anymore.

Maybe now it is time to come to the end of your career—Abschied von der Wissenschaft.[22] *Sometime ago you retired and we hear that you have really withdrawn from science.*

That is a part of my way of doing it. I am a person who can change rapidly. There is a time for everything. There is a time to be young; there is a time to work and to

[22] Departing from science.

travel and there is a time to retire when you have deserved it. There are a lot of young people who are looking to do the next great thing. Why should we not quit one day and enjoy the life.

Nowadays you are no longer working in science?

I am not working on scientific problems. That is true. I am still interested in what is going on in oceanography and climate research.
Your last paper is written?

Is written in 1993, quite a while ago.[23]

II

Let's talk about changing themes, the effect of new methods and opportunities, experiments, models, remote sensing.

Some of it we have already touched. The big subjects, that I just mentioned like ocean monitoring, free data exchange, and so on—these are problems that science faces and that have to be solved beside the scientific problems. When it came to ocean monitoring, there are always new things—for instance, during my lifetime the satellites came up. I was one of the members of the initial TOPEX committee that Karl Wunsch started up. We were discussing and were very, very excited about the possibility of monitoring global sea level variability in areas without islands or fixed observation points. That is of course a step into the future of oceanography. The continuous observation of our environment is an enormous step forward.

Could you try to describe what the big topic in the forties was, in the fifties and so forth? We just go through these six decades and you try to outline what to you was of most interest or significance.

This is a good way to start. Before the World War deep ocean circulation was the interesting stuff, Defant and Wüst and Sverdrup. In the 40s, I cannot really tell you. In the 50s it was surely the ocean eddies.

Science is per se a matter of fashion. When I was a student, every physicist had to study atomic physics, and if you were studying acoustic or anything else, you were second-rate.

So, in this sense I am asking for the fashions, wie lang waren die Röcke, die wissenschaftlichen,[24] in the 70s?

The eddies . . . of course, biochemical cycles.

[23] Wyrtki, K., 1993: Global sea level rise. Proc. Circum-Pacific Int. Symp. Earth Environment, National Fisheries Univ. Pusan, Pusan. D. Kim and Y. Kim, Eds., 215–226.

[24] How long were the skirts in science?

Already in the fifties or sixties?

GEOSECS—seventies. The eddies were the first big problem after the war. I don't think the eddies started in the 50s, definitely in the 60s.

Once I got a student who wanted to make a Ph.D. Peter Duncan came from South Africa and he brought along the results of one cruise that they made to the southwest of Africa and I did nothing but apply another principle of Wüst. If you have observations, which haven't been used yet, you write a paper about it. I made him immediately write a paper about an eddy in the subtropical convergence south of South Africa. He wrote that paper in ten days and it was accepted by JGR. The background was a frivolous statement I had made in a class: any graduate student can write a paper that will be accepted by JGR and I gave them the recipe: New observations that haven't been published, a straightforward analysis, no controversial statements, 4 pages, 3 illustrations.

Four pages text?

Yes, at most. Today I would say two pages.

Why was the interest in eddies so big? One thing of course, it was possible to observe eddies; on the other side, could you already estimate what the role, what the importance of eddies in the general climate dynamics is?

That was more ocean dynamics than climate dynamics. People thought that a better knowledge of ocean eddies would explain the energy dissipation in the western boundary currents and in ocean circulation in general, because all ocean circulation theories were dependent on dissipation.

So, that was the time of the fifties and sixties.

These were the 50s, 60s, early 70s.

Then came the International Decade of Ocean Exploration. All these big projects were started in the seventies: the biochemical cycles, Antarctica, the Drake Passage, there was NORPAX, which was the project I joined in.

Is it fair to say that before the war people were interested more in the deep ocean circulation and the overall picture and after the war more in processes and in case studies on eddies. In the seventies it was the phase of integration, so that the people were more interested in longer observations, in variability. Is that right?

You can say so. Actually NORPAX was the first big project that studied ocean–atmosphere interaction. The database became sufficient to look at a larger picture—that means how an ocean affects a continent.

Was Namias very important in this respect?

Of course, I had a very close relationship with Namias. When you ask for people: there were of course Bjerknes and Namias. We were together very often in meetings and had many long discussions.

Was he approximately your age?

Namias was 14 years older, he died in 1997, and Bjerknes was much older—he could have been my father. This was a time of enormous cross fertilization.

What about the nineties?

The nineties are clearly climate, the chemical and biological cycles in the climate system. These are the next big topics, not the physical cycles of climate.

Could you say something on the role of experiments? Like in GARP when people came together to make a big effort to observe the atmosphere or the ocean or the boundary layers, intensively for a limited time, and then go back into the laboratories?

Experiments are absolutely necessary. Experiments are the basis of physics. We do process experiments, which are real physics in the ocean, where you try to learn . . .

Could you give an example?

Such as, how does the Ekman layer work? These are physics experiments. But then you have to make other experiments and these are very often not recognized as experiments: is Global Change an experiment? Now, you see, one important thing about geophysics is, and I tried to explain that to my students, physics is based on experiments where you can control one factor at a time. But in geophysics all factors are changing simultaneously. Nature is making experiments for us and as geophysicists we are very often simply put in the role of the observers. We can't control the experiment. If you make an experiment on a hurricane, you don't control the experiment.

How did you feel the assistance of numerical models, which was increasing with time? Numerical experiments . . .

Numerical experiments, you mean models. Models are an essential part of physics and sciences today. There is no question about that. You need models for everything. You only have to use them in the right way. There are many different kinds of models. Models are to simulate certain physical processes. But a model is an approximation that can be used to study physical processes. Then there are models that predict weather. They have limitations. What are the limitations? In prediction it is chaos and turbulence. Then you make models of the tides. They are probably very good, because they have solid physics behind it and the process is truly repeatable, because it is forced. Then you can make models that are plainly speculative, that means, where we are trying ideas. The question is what you make with models. There is nothing wrong with models, but how you interpret the model, that's the important item.

Could you just give an example of a speculative model?

I would say, modeling 100 years of climate change is speculative.

You are not talking about models like Stommel's?

No, Stommel's model is a conceptional model of a process in which he explores the effect of β. He explains.

During your scientific career the role of models must have changed. I guess, when you were in Kiel there were no models.

There were of course Stommel's models. Conceptual models have always been part of physics. And experimental models also.
Then came computer models and took more and more part of the science. How did you experience that?

With a certain amount of skepticism, but the same skepticism I would have to an experiment. That means, I don't challenge the model, but the conclusions that people draw from the model.

Go back to basic physics. An astronomer makes an observation, first he speculates what happens. That is the first step, may be right, may be wrong. Then he makes a theory. Mathematically a model is equivalent to a theory. Then he asks what goes into the theory? What are the basic assumptions of the model?

I have a question to you. The big ocean circulation models that we are having today and that show many details of ocean circulation, do they include tides?

We[25] *have in the meantime a circulation model, which includes tides.*

For the world ocean? Do the tides interact with the circulation? They must interact. They cause mixing and will dissipate energy.

Yes, for the world ocean. The tides interact in this model. Normally the circulation models do not include tides.

When somebody comes with a result on ocean circulation you ask what does it resolve? What's the effect of tides? You can put a mixing parameter in your models which specifies a number. But how good is that? These are the challenges to the models.

But you said you would challenge the modeler, not the models.

The conclusions from the models.

Are there some systematic problems with models?

No, I don't think so. Models are part of physics, but you have to be skeptical about the results. Models are as much part of physics as experiments are. They are only a different way of conducting experiments. Don't misunderstand that.

Another thing which came up in your career was remote sensing. Suddenly there were satellites and you could observe the whole world from space. What did they change?

Well, again a personal approach. If I want my appendix out, I hire a doctor, if I want to compute I hire a computer programmer, and if I want to do engineering I hire a competent engineer. I don't do these things myself. That is simply my approach to

[25] Statement by Jürgen Sündermann.

the satellites. There were other people there who did it much better than I would have done it.

Did the advent of satellites change your science?

Oh yes, it has changed. It began with surface temperatures. That was the first parameter for which we got global coverage. Then came the clouds, cloud motion vectors, that gave us the winds. This was an enormous advance.

What about sea level elevations?

Eventually TOPEX and the altimeters. I did not participate in the use of altimeters anymore. We had younger people who were doing an excellent job at that. It is not necessary that you do everything.

III

In 1948 the theories about the westward intensification of the big gyres were published.

I was a student at that time and I remember that Wüst showed me the paper by Stommel and it was a big surprise and everybody thought that it is a wonderful thing that happened. So, these insights are being recognized when they happen.

The physics behind the β-effect and the driving by the wind is relatively simple. Why has it not been detected earlier?

Because nobody had the idea. That is the reason.

What are the causes of scientific progress?

All the four points you put here.[26] Gelegenheit ist Zufall,[27] it is certainly not planned. The progress in science I don't think is planned. It happens when certain problems are ripe for a solution.

What is the role of nations?

Well, we can keep that short. First of all the role of the various nations in ocean research is basically dependent on their wealth. The wealthy nations can put a lot of effort into research and they will succeed because research after all is expensive. I don't really know what to answer to that. Different nations are definitely interested in different things. Japan, for instance, is very interested in resources in fishery and so on. Other nations are interested in other aspects such as oil or geology.

Such as . . . military?

Military is of course an option. Russia and the U.S. have been tremendously interested in military aspects of oceanography.

---

[26] On the tentative list of questions, the items funding, opportunity, people, and coincidence were listed.

[27] Opportunity is coincidence.

Nations can also act in the opposite way. This is what I want to point out with regard to Indonesia. You know, when Arnold Gordon planned this big throughflow experiment, Fritz Schott wanted to do the moorings, Arnold Gordon the hydrography, and I came in a little bit with sea level, but the Indonesians didn't want international participation. I remember one international meeting on which an Indonesian admiral said flatly "we don't want any damned foreign ship in our waters." So Indonesia has excluded to a large extent progress because they did not allow other nations to come in and work with them. And this is why it has remained a desert of knowledge.

What about international organizations?

International organizations are necessary, in order to get ships into foreign waters, to make data exchange and similar things, to enable international cooperation, because you can't install observing stations somewhere unless you have permission of that country. You can't do research within the 200-mile zone unless you cooperate with that country. All these international organizations are necessary. Some do very good jobs, some not. But there is a need for it.

What about physical oceanography as part of a more general environmental science?

Physical oceanography is in some way basic to all the other branches of oceanography, because all the others are simply embedded in the physical environment. In order for biologists, chemists in particular, to explain their results, they have to go back to ocean circulation and to physical processes. For that reason it will always be the main part of oceanography. Maybe not the most important one, but the main indispensable part. You cannot explain plankton distributions and productivity without knowing about circulation, mixing, and other processes.

Has oceanography become also a sub-discipline of climate research, or global change research?

Oceanography exists quite independently of climate research. It is certainly not a subdiscipline, but a very important component of it, because of ocean–atmosphere interaction. The ocean definitely plays more than the role of a copper plate

A wet copper plate.

Yes, something like that. The ocean is awfully active. The ocean is handling the storage of heat. When it comes to climate prediction or long-term weather prediction, then the ocean plays a major role in providing the heat storage and in advecting heat. Advection is a much-neglected phenomenon in most studies or explanations of the ocean–atmosphere system.

What was the background of you mentioning the copper plate? Were there people who said the ocean is just a copper plate providing heat for the atmosphere?

This claim has been made by some meteorologists. It has seriously been claimed that the ocean doesn't count, but we are beyond that now.

Could you say names of proponents?

I would say GFDL.

Should the physical oceanographers give more interest to the other disciplines, to biology, to chemistry, in order to give more exact explanations into these sciences?

Oh yes, it doesn't hurt, there will always be physicists who are just physicists, but for an oceanographer general knowledge of the surrounding fields of interest is very important, if he wants to make his knowledge applicable. If he wants to talk with a plankton man about vertical mixing or such things, then it is very important that he has understanding of the mutual subject. So I would say it is a general principle: additional knowledge doesn't hurt.

What does it mean for the education of the students? Should we still have this classical education that they study physics, mathematics, and so on? Or should we have some general education in marine sciences?

It should not be mandatory but it should be very much encouraged. To make things mandatory is not a good idea. That means you would prevent a computer programmer to become an oceanographer by forcing him to do some biology in which he is not interested at all.

In your career there was always some link to applications. When you did the tuna business, when you were in Indonesia, there was always an element of usefulness. Is that so?

No, not useful, but realistic. I'm a realist and I want to work on things that represent the real world that give an understanding of what there is. I am not a friend of speculations and fancy theories, I like to analyze facts and put them together and explain them.

Did you have to write in your proposals "this is important for fisheries or for..."?

You usually say that.

On your list of items, you ask about the role of science organizations, big science, universities, centralization. Big projects are necessary, for the very simple, pragmatic reason that an individual can't do them. An individual cannot launch a satellite and use all the data that come back. For big experiments you need cooperation of many people. This is a practical question. But big science does not mean that one should take the funding away from all the individual scientists. Individuals have their own ideas and often very good ones. There are enough scientists that don't like to be involved in community projects. So one has to keep a balance between them. The same basically applies to universities versus government organizations. The universities are providing diversity and individualists. They allow the individual scientists to do work outside the mass, and they give him the freedom to do what he likes to do. In contrast, government science is mostly directed science, that means the people involved in it are being told what they have to do.

But there are also research institutes like Max-Planck-Institutes.

They are taking a middle position between the two. Depending on the country, some of these research institutes are tending more to be like university institutes, others more like government institutes. So, there is a real spectrum between a concerted government effort by the Navy and a small university with individuals. The whole spectrum exists, and any part of the spectrum is useful.

When you came to Hawaii in 1964, the Department of Oceanography had just been established. You were among the first professors of that department. The department grew relatively quickly over, say, 25 years and then we made this new school.[28] *So the number of colleagues grew tremendously. How has this growth influenced your work as a professor, as a teacher?*

I personally prefer to be in a small university, in a small institute that is relatively independent. I do see the need for bigger organizations, but there is enough good science coming out of small institutions and individual efforts as there is out of big institutions. The growth did not at all affect myself—I was in a position to remain sufficiently independent from the big institution to do what I wanted to do. This may not be the case for all scientists in that institute.

How efficient is the steering through soft money projects? When the government is saying they want to support a certain type of research and they offer soft money.

They have said that many times to me and I had to say, "No, thank you". One day the Office of Naval Research representative told me "climate is out. Forget climate funding, anything climate related." I said "Fine. What can I do, I go to the next agency".

Is it not a very efficient type of control, which is exerted by the government?

No, the agencies have their own priorities and there is a good reason for that. The Navy has certain priorities, they can't just support the Honolulu symphony.

You had sufficient sponsoring organizations to get money for any idea you would like to realize?

Yes, you are right. We have been in the U.S. in the fortunate situation that we had over decades surplus funding—my opinion. We have enough funding to keep all the good scientists busy. There will always be people who say "I should get funded." No doubt about that. There are always people who say funding is not enough.

Big projects. There are certain things for which the big projects are necessary. The weather service can't live without big projects. And the fishery service also cannot. But this is applied science, this is in some way even technology, but when it goes beyond that and it comes on the National Science Foundation level, then the peer

---

[28] School of Ocean, Earth Science and Technology (SOEST).

review system works well and there should be no centralization. I am not much in favor of these centralized projects. I've been for many years chairman of NORPAX. It was really not that centralized, but nonetheless funding was in some way restricted to the program.

A new term I like to bring in is "political science". When politicians use science it gets hairy. There is a story being told in recent months that a government scientist and a government official were talking with each other and the government scientist said, "oh, my data show this" and the government official said, "why don't you change the data?" That is "political science". And that's what scientists should avoid.

Is this a real problem in the United States now, or worldwide?

It is a real problem for all countries, if politicians want to tell their population something that is contrary to scientific evidence. In industry, this situation has existed for a long time, but it becomes dangerous to scientific freedom if such situations would happen.

What is the influence of media and the impact of media attention that certain people receive?

Media attention is good for science but media attention very often confuses the issues, because they might very well get practical and political aspects into it.

Another problem is "truth in science". In this case you have to differentiate between science and scientists. Science per se eventually converges on the truth. We learn things and they become knowledge. Scientists are not necessarily very objective when it comes to make propaganda for a cause, like the blown-up predictions that are now being made of weather and climate, of El Niño in particular. We are hearing predictions that are being blown-up by the press and of scientists making statements, which they cannot defend in the long run. This is dangerous for science.

Why do they make these statements?

Because they are human. They want to show off. If you stand before a TV camera, you give a big talk, you say El Niño is coming....

What do you think about present-day forecasts of El Niño and La Niña? How good are they, for how long are they good?

Scientists like to make forecasts. Forecasts are made about the weather and we know reasonably well what the limitations are. Forecasts of climate are a lot more uncertain and in particular El Niño forecasts. There are several models on El Niño. If seven forecasters are making an El Niño forecast, then four may be correct, three may be not correct. The four who are correct claim in front of the TV camera that it was a success, the three who were incorrect are being quiet until the next time. Most forecasters—I could show you examples—are saying after the fact that they made a valid forecast.

Then they say they have made a forecast nine months in advance. The question is what did they forecast? Did they forecast the beginning of El Niño or the peak of El Niño? You will find out that they forecasted the peak of El Niño, which was, say, in August. The El Niño started in March and they made the forecast in December. December to August is nine months so they claim they made a nine-month forecast, when actually they made only a three-month forecast.

When you make a forecast, you have to be awfully specific what you are forecasting, and not just make a press release that something will happen. Therefore, I am quite skeptical about these forecasts. I had a nice email exchange with my friend Glantz in Boulder—he is an expert on social-economic impacts of El Niño and he would like to use forecasts to tell the farmers what they have to do, to seed rice or cotton, for example. He asked whether the last El Niño has been forecast and he came to the conclusion "not really". When El Niño started, when the first indications came up, people started to claim that they had forecasted it.

There should be a better control about what El Niño forecasts are made. And scientists should be a lot more honest.

Is it time for one big international center, such as the European Center of Medium Range Weather Forecast, for El Niño forecasting?

Yes, it may be necessary and economical to have a center that collects all the data because the data collecting effort would be common to all. Making a forecast is the use of the data. That comes one step afterwards, and can be made on the same data by many different people.

The success of the European Center of Medium Range Weather Forecast is based on their data collection and data analysis processes.

And then you give the data to the forecaster in Moscow, Frankfurt, or elsewhere. And the forecaster makes his particular forecast for a region that he knows better than the others. Of course the computer models may spit out the same information. In the end one best model may develop. We are at the beginning of the era of models. There are great things to come.

IV

The role of your colleagues, of the working team, of schools. Did you experience during your scientific career that there are existing schools, groups which have certain minds, certain theories, is this important in oceanography?

The exchange of ideas, opinions, plans, and so on is most important for a scientist. Otherwise you become very soon sterile. It happens on large scales, through conferences in an objective way, through personal friendships most consistently, and most scientists participate in this interaction.

Die lieben Kollegen[29] come of course in all sizes and shapes. There are the nice ones, the ones that are generous, that are stimulating, and that are open-minded—Hank Stommel was a prime example of that. And then the average that doesn't care and is uninterested or irrelevant to you. Then of course the bad guys, the people that are arrogant, troublemakers and are vicious. You have them all, scientists are just like any other people.

You essentially select a group with which you feel comfortable and want to do things. That group changes with time, with the interests that you have. Some people stay a whole life in the same group because they never get away from a particular subject. You change the groups when you change topics; you talk to other people when you deal with deep circulation than when you do El Niño or climate.

Are there different ways of thinking? Is there an American way of thinking in oceanography, or a western European or a Russian way of thinking?

There will always be schools, that means interest groupings around a problem like NORPAX or like GEOSECS. GEOSECS was one of the closest groups that I have ever seen in scientific cooperation. There are more loose groups, but it is hard to say—I haven't been too much involved in group efforts.

Have you experience that certain groups were blocking progress?

Oh yes. As already said, science is very often a matter of fashion. When everybody was in ocean eddies, we had to fight long battles to get ocean monitoring going. In later years the people who wanted to make so-called process-oriented experiments were fighting bitter battles at the National Science Foundation with other people who wanted to make ocean surveys like GEOSECS or like the WOCE sections.

In the sixties there have been longstanding battles between the U.S. East Coast and West Coast, Woods Hole versus Scripps. That went on. It was a competition of opinions, very often. The Woods Hole people were interested in controlled experiments like MODE and POLYMODE and the kind, and the Scripps people were largely interested in the larger ocean surveys that had relation to fisheries, climate, and to largescale features. These are opinions that go back and forth. There is fashion in science and group building, no doubt.

V

What are your forecasts of the future of science?

My general forecast of what will happen in the future is that first of all we will get truly global coverage of observations, from satellite and eventually from other systems like

[29] The dear colleagues.

the TOGA TAO and similar systems, because the satellites don't penetrate inside the ocean.

So far the Southern Hemisphere is grossly neglected. The Southern Hemisphere will be in the end more decisive for the interpretation of climate change than the Northern Hemisphere, because it connects the three oceans, and it is the most powerful ocean–atmosphere engine that we have and it has not been sufficiently studied because of the lack of data. People study these things first when they have good data.

No wonder that certain people in the sixties did not want to go to Antarctica because they became seasick and found it too cold.[30]

You are so right about that. But there are other people who love it.

Could you make another kind of forecast, not about science, but about the nature itself? Within the next fifty years, will there be global warming? How will the average temperature at the sea surface change within the next fifty years?

There are many people working on that problem. I have only an opinion. We will see a continuation of global warming, whereby I am not quite positive whether it is primarily natural, or primarily man-induced. Probably both components are important. When you ask me how big that change will be over fifty years, I would say, not more than it has been in the last fifty years. With regard to sea level my successor[31] in the sea level project has made a very interesting plot. It starts with the first prediction of 3 m over one century, or something like that by the club of Rome. Then came a few years later 1 to 1.5 m and then 0.2 to 1 m and later 40 cm. He put a regression line through that cloud of dots which has an exponential decay to the average value of the last 100 years. So this is where the forecasts go. They converge towards the extrapolation of the last 100 years. That is approximately correct for the next 50 years, 10 cm in 50 years, which is a little more than in the last century, which was 15 cm.

Climate change will always be of interest, ocean–atmosphere interaction in connection with climate change. It will lose in importance. What will gain in importance will be chemical pollution, biological change—which is of course embedded into climate change—water resources. Years ago in Cabo San Lucas in Mexico, I had to spend three dollars for a liter of water. I said to my friends, "before you die you will see that water is more expensive than gas".

But water resources have a lot to do with climate change.

They do, they are a fundamental part of climate change. But no doubt, the warmer, the more rain you get. It may not fall at the right places. But basically it will still fall in the same places as now. There may be shifts, but unless we get a total change of atmosphere circulation the monsoons will always happen.

---

[30] See page 209.

[31] Gary Mitchum.

Do you think that independently of climate change we are running out of water?

Yes, I think so. It will be a scarce commodity.

Have you anything to do with paleoclimatology?

No, I shied away from it intentionally because to me it was too speculative.

Do you think it will play an important role in the future?

Any part of science that can be thoroughly documented is important.

Will it become fashion?

It has been a fashion. If it will remain a fashion, this is another thing. I don't have an opinion on that. It lends itself to lots of speculations and hypotheses, because it is so difficult to prove anything.

There are many established facts about paleoclimate. No doubt that we know a lot about the Ice Ages. That is beyond speculation, but if you start to link Ice Ages and ocean circulation you get into speculation.

Do you believe in these results indicating sudden climate changes?

It depends what you call sudden.

Within decades of years.

Decades it seems to be a little fast. Hundred years I would say is perfectly possible. But this is again just an opinion. In order to get climate changes you have to start substantial melting processes or accumulation processes and they do not happen in decades.

When we have what you call truly global monitoring systems, will we get long-range forecasts with models based on the good knowledge of the dynamical state of the ocean?

What I said before—models are in their early stages of development. That means we will get many more surprises out of models, we will get much, much better models in the future. I am talking about climate models, not necessarily applied models like ship routing or so. Better and more comprehensive observations will feed better information to models. I don't know to what extent the physics of the models need to be improved, but I think they will be. Science doesn't give up on these things, there is always something that can be done better. Our understanding of the processes, for instance the basic process of ocean–atmosphere interaction, that govern nature will increase, and therefore the models will improve.

But there are limits to predictability. Many scientists and certainly many outsiders do not want to accept this. People always want to have a certainty about a prediction. They think if somebody gives them a prediction it should be certain. But this is by no means so. A correlation of seventy percent means that two times you are right and one time you are wrong, roughly speaking. So if you make forecasts that go beyond the dynamical range of the model where turbulence or chaos takes over your forecast becomes essentially statistical. You can run 25 models 100 times each and you have 2500 predictions and you average that and you think you have made a forecast. No, because only one will be realized by nature. Nature will not realize the average. There is a limit to forecasting.

Another technique of forecast is basically the extrapolation; actually, it is more than an extrapolation, for instance, when you predict climate, you are projecting into the future. This is better than an extrapolation. You are projecting what developments or what changes can go on and you may give a certain envelope to this projection. The envelope will become wider and wider with time. These things are all recognized by reasonable scientists. I don't say anything new.

Do you expect new developments or breakthroughs by new instruments?

I have too little knowledge about instruments. The satellites are new instruments, if you want to say so. We will see more.

The basic principle of Dr. Wyrtki is, if you look closer at something with a new instrument you find something.

That's what Wüst said and I demonstrated it.

Will there still be interest in science in fifty years? Will people listen to scientists?

There will always be curiosity, science is driven by curiosity. There are always people who are curious about things and they want to know it better.

We haven't finished the prediction. You ought to look at developments that in the future may take place. One point that is totally unknown to me is warfare, fortunately. We do not have the slightest idea what the role of oceanography will be. It has had a considerable role in the last thirty years. More money has definitely gone into antisubmarine warfare than into academic research. The other open problem is of course the population explosion and what to do about it. These problems will entertain us in the next fifty years.

You wrote about that. I remember you had an article when you discussed the prospects of climate change.

I said that sea level rise will be a picnic compared with the population explosion.[32]

VI

You have already spoken a bit about what you consider your most important achievements. You said freedom of data, the monitoring idea, and other things. Is there anything else you would say which has been a major achievement of yourselves?

The other items are plainly scientific ones. There is of course El Niño; its explanation as the ocean response to the atmosphere and later on the explanation of the El Niño cycle

---

[32] Wyrtki, K., 1989: Sea level rise—the facts and the future. *Pac. Sci.*, **44** (1), 1–16.

as an accumulation of warm water that eventually changes atmospheric circulation and triggers the next event.

Which are your favorite own publications?

These are the thermohaline circulation from 1961[33], and the deep sea basins, the oxygen minima from 1962[34]. Then I would mention the Peru current, which linked the horizontal and vertical movement in a very large area of the ocean.[35] Then you have the Indian of the Ocean Atlas and the analyses of the Indian Ocean circulation and with that came the Indian Ocean jet.[36]

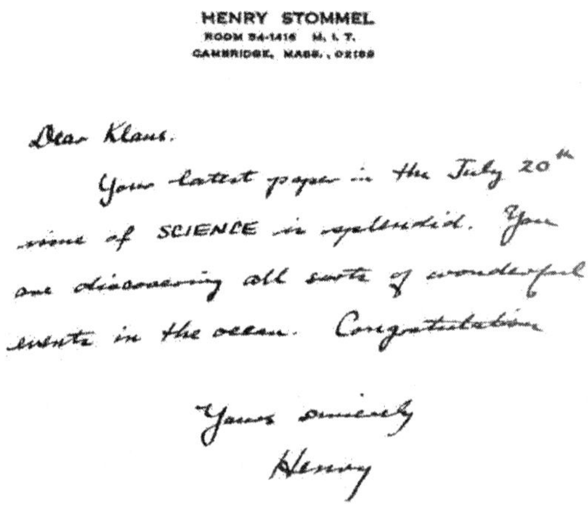

HENRY STOMMEL
ROOM 54-1416 M.I.T.
CAMBRIDGE, MASS., 02139

Dear Klaus.

Your latest paper in the July 20th issue of SCIENCE is splendid. You are discovering all sorts of wonderful events in the ocean. Congratulations

Yours sincerely
Henry

Letter from Hank Stommel

From Wyrtki, 1963, Bull. Scripps Inst. Oceanogr.

What about your Baltic studies?

The Baltic study was an important piece of work for me, it was an effort to understand the water budget of a small sea that has sufficient information, and to understand both the annual cycle of exchange and the fact that this annual cycle was basically wind driven.[37]

---

[33] Wyrtki, K., 1961: The thermohaline circulation in relation to general circulation in the oceans. *Deep-Sea Res.*, **8** (1), 39–64.

[34] Wyrtki, K., 1962: The oxygen minima in relation to ocean circulation. *Deep-Sea Res.*, **9**, 11–23.

[35] Wyrtki, K., 1963. The horizontal and vertical field of motion in the Peru Current. *Bull. Scripps Inst. Oceanogr.* Univ. Calif., **8** (4), 313–346.

[36] Wyrtki, K., 1973: An equatorial jet in the Indian Ocean. *Science*, **181**, 262–264.

[37] Wyrtki, K., 1954: Schwankungen im Wasserhaushalt der Ostsee. *Dtsch. Hydrogr. Z.*, **7** (3/4), 91–129.

Is the Baltic a model of the global ocean?

In some ways, yes. It has a wind-driven exchange, the Baltic is either pushing water out or holding water in, depending on the weather. The study about the water balance of the Baltic basically summarized the whole story. The Fehmarn belt papers were about the dynamics of the exchange. [38]

Then afterwards the El Niño papers, and finally sea level and of course all the things that had to do with the dynamics of the Pacific upper ocean.

Sometimes people say scientists are creative when they are 25/30 years. Then, after that the creativity is declining. Is that so in your view?

That is putting it too early. Our typical Ph.D. age is 30 now. I was 25. But even at that time it was an exception, it was more like 27 or so. Unless you make an exceptional discovery as a graduate student, you start to be a scientist by 30. You need a buildup time of maybe 10 years. Between 40 and 50 you should have your peak productivity in new things. Between 50 and 60 should be a period where you consolidate knowledge and integrate.

Have you thought of writing a book?

Yes, I have. What came nearest to a book was the NAGA Report[39] which you may call a monograph, also the Indian Ocean Atlas[40] is a big piece of work. I intended to write a book with the title *The Water Masses and Circulation of the Indian Ocean* and I gave it up since it takes about five to six years to write and by that time much of the information is superceded by new knowledge. Knowledge is accumulating these days at a rate that you can say after a decade things are old. That's a too short a lifetime for a book.

You always had interest not only in science but you traveled a lot and you enjoyed also the nice environment here in Hawaii. To what extent was this part of your life also important for the science? This mixing of more private life and scientific life.

It was a very lucky and favorable choice. First of all it was a true choice to come to Hawaii. After I had been here in 1958 for the first time I decided essentially that I would like to live here. Then it was the opportunity that a new institute was being built up in the middle of the Pacific.

---

[38] Wyrtki, K., 1953: Die Dynamik der Wasserbewegungen im Fehmarnbelt I. *Kiel. Meeresforsch.*, **9** (2), 155–170.

Wyrtki, K., 1954: Die Dynamik der Wasserbewegungen im Fehmarnbelt II. *Kiel. Meeresforsch.*, **10** (2), 162–181.

[39] Wyrtki, K., 1961: Physical oceanography of the southeast Asian waters. Univ. Calif., NAGA Rept., No. 2, 195 pp.

[40] Wyrtki, K., 1971: *Oceanographic Atlas of the International Indian Ocean Expedition.* National Science Foundation Publication, OCE/NSF 86-00-001,Washington, DC, 531 pp.

We have to come to a conclusion... the tape is ending.

I have no regrets about the things I have done. I have enjoyed the scientific career that I have made. I would do the same thing, it may not turn out the same way because we are subject to chance, you know, but basically I would do the same.

The interview was conducted on 25 February 1999 in two sessions of about 2 hours each – the first in Peter Müller's lab in the Marine Science Building in Honolulu, the second in Klaus Wyrtki's apartment. Participants were Hans von Storch, Jürgen Sündermann, and Lorenz Magaard. Two tapes were transcribed by Ilona Liesner and edited by Klaus Wyrtki and Hans von Storch.

PUBLICATIONS

Wyrtki, K., 1950: Über die Verteilung der Trübung in den Wassermassen der Beltsee und ihren Zusammenhang mit den hydrographischen Faktoren. Ph.D. dissertation, Univ. Kiel, FRG, 49 pp.

Wyrtki, K., 1950: Über die Beziehungen zwischen Trübung und ozeanographischem Aufbau. *Kiel. Meeresforsch.*, **7** (2), 87–07.

Wyrtki, K., 1951: Der Einfluß des Windes auf die Wasserbewegungen durch die Straße von Dover. *Dtsch. Hydrogr. Z.*, **5** (1), 21–27.

Dietrich, G., K. Wyrtki, J. N. Carruthers, A. L. Lawford, and H. C. Parmenter, 1952: *Wind conditions over the seas around Britain during the period 1900–1949.* German Hydrogr. Institute, Hamburg, **8** (52), 283, 38 pp.

Wyrtki, K., 1953: Untersuchung der Strömungsverhaltnisse im Bereich der Neuen Schleusen in Kiel Holtenau. *Berichte Kanalbauamt*, 19 pp.

Wyrtki, K., 1952: Der Einfluß des Windes auf den mittleren Wasserstand der Nordsee und ihren Wasserhaushalt. *Dtsch. Hydrogr. Z.*, **5** (5/6), 245–252.

Wyrtki, K., 1953: Die Dynamik der Wasserbewegungen im Fehmarnbelt I. *Kiel. Meeresforsch.*, **9** (2), 155–170.

Wyrtki, K., 1953: Ergebnisse über die Verteilung der Trübung in Küstennähe. *Veröff. Inst. Meeresforsch. Bremerhaven*, **2**, 269–278.

Wyrtki, K., 1953: Die Bilanz des Längstransportes in der Brandungszone. *Dtsch. Hydrogr. Z.*, **6** (2), 65–76.

Wyrtki, K., 1954: Der große Salzeinbruch in die Ostee im November und Dezember 1951. *Kiel. Meeresforsch.*, **10** (1), 19–24.

Wyrtki, K., 1954: Die Dynamik der Wasserbewegungen im Fehmarnbelt II. *Kiel. Meeresforsch.*, **10** (2), 162–181.

Wyrtki, K., 1954: Schwankungen im Wasserhaushalt der Ostsee. *Dtsch. Hydrogr. Z.*, **7** (3/4), 91–129.

Wyrtki, K., 1956: Monthly charts of surface salinity in Indonesian and adjacent waters. *J. Conseil*, **21** (3), 268–279.

Wyrtki, K., 1956: The subtropical lower water between the Philippines and Irian (New Guinea). *Mar. Res. Indones.*, **1**, 21–52.

Wyrtki, K., 1956: The rainfall over the Indonesian waters. Lembaga Meteorologi dan Geofisik, Verhandelingen No. 49, 24 pp.

Wyrtki, K., 1956: The computation of oceanic and meteorological fields of motion with friction proportional to the velocity. *Mar. Res. Indones.*, **2**, 1–26.

Wyrtki, K., 1957: Die Zirkulation an der Oberfläche der Südostasiatischen Gewässer. *Dtsch. Hydrogr. Z.*, **10** (1), 1–13.

Wyrtki, K., 1957: Precipitation, evaporation and energy exchange at the surface of the southeast Asian waters. *Mar. Res. Indones.*, **3**, 7–40.
Wyrtki, K., 1958: The water exchange between the Pacific and the Indian Oceans in relation to upwelling processes. Inst. Mar. Res., Djarkata. *Proc. Ninth Pac. Sci. Congr.*, **16**, 61–65.
Wyrtki, K., 1960: Surface circulation in the Coral and Tasman Seas. Div. Fish. Oceanogr., Tech. Pap. No. 8, 44 pp.
Wyrtki, K., 1960: On the presentation of ocean surface currents. *Int. Hydrogr. Rev.*, **37** (1), 153–174.
Wyrtki, K., 1960: The Antarctic convergence—and divergence. *Nature*, **187** (4737), 581–582.
Wyrtki, K., 1960: The Antarctic circumpolar current and the Antarctic polar front. *Dtsch. Hydrogr. Z.*, **13** (4), 153–174.
Wyrtki, K., 1961: The thermohaline circulation in relation to general circulation in the oceans. *Deep-Sea Res.*, **8** (1), 39–64.
Wyrtki, K., 1961: The flow of water into the deep sea basins of the western South Pacific Ocean. *Aust. J. Mar. Freshwater for Res.*, **12** (1), 1–16.
Wyrtki, K., 1961: Optical measurements in the Coral and Solomon Seas. Symp. on Radiant Energy in the Sea, *Int. Union Geod. Geophys.*, Monogr. No. 10, 51–59.
Wyrtki, K., 1961: Physical oceanography of the southeast Asian waters. Univ. Calif., NAGA Rept., No. 2, 195 pp.
Wyrtki, K., 1962: The oxygen minima in relation to ocean circulation. *Deep-Sea Res.*, **9**, 11–23.
Wyrtki, K., 1962: The subsurface water masses in the western South Pacific Ocean. *Aust. J. Mar. Freshwater Res.*, **13** (1), 18–47.
Wyrtki, K., 1962: Geopotential topographies and associated circulation in the southeastern Indian Ocean. *Aust. J. Mar. Freshwater Res.*, **13** (1), 1–17.
Wyrtki, K., 1962. Geopotential topographies and associated circulation in the western South Pacific Ocean. *Aust. J. Mar. Freshwater Res.*, **13** (3), 89–105.
Wyrtki, K., 1962: The upwelling in the region between Java and Australia during the southeast monsoon. *Aust. J. Mar. Freshwater Res.*, **13** (3), 217–225.
Wyrtki, K., 1963. The horizontal and vertical field of motion in the Peru Current. *Bull. Scripps Inst. Oceanogr.* Univ. Calif., **8** (4), 313–346.
Wyrtki, K., and E. Bennett, 1963: Vertical eddy viscosity in the Pacific equatorial undercurrent. *Deep-Sea Res.*, **10** (4), 449–455.
Wyrtki, K., 1964: Total integrated mass transports and actual circulation in the eastern South Pacific Ocean. *Studies in Oceanography*, Japan, 47–52.
Wyrtki, K., 1964: Upwelling in the Costa Rica Dome. *Fish. Bull.*, **63** (2), 355–372.
Wyrtki, K., 1964: Surface currents of the eastern tropical Pacific Ocean. *Inter-Amer. Trop. Tuna Comm. Bull.*, **9** (5), 270–304.
Wyrtki, K., 1964: The thermal structure of the Eastern Pacific Ocean. *Dtsch. Hydrogr. Z.*, Ergänzungsheft A, **6**, 84 pp.
Wyrtki, K., 1965: Summary of the physical oceanography of the Eastern Pacific Ocean. Univ. Calif. IMR Ref. 65-10, UCSD-34P99-11, 69 pp.
Wyrtki, K., 1965: The annual and semiannual variation of sea surface temperature in the North Pacific Ocean. *J. Limnol. Oceanogr.*, **10** (3), 307–313.
Wyrtki, K., 1965: The average annual heat balance of the North Pacific Ocean and its relation to ocean circulation. *J. Geophys. Res.*, **70** (18), 4547–4559.
Wyrtki, K., 1966: Seasonal variation of heat exchange and surface temperature in the North Pacific Ocean. Univ. Hawaii, Tech. Rep. HIG-66-3, 8 pp. + 72 figs.
Wyrtki, K., 1966: Oceanography of the eastern equatorial Pacific Ocean. *Oceanogr. Mar. Biol. Ann. Rev.*, **4**, 33–68.
Wyrtki, K., and V. Graefe, 1967: Approach of tides to the Hawaiian Islands. *J. Geophys. Res.*, **72** (8), 2069–2071.

Wyrtki, K., and R. Kendall, 1967: Transports of the Pacific equatorial countercurrent. *J. Geophys. Res.*, **72** (8), 2073-2076.
Wyrtki, K., 1967: Circulation and water masses in the eastern equatorial Pacific Ocean. *Int. J. Oceanol. Limnol.*, **1** (2), 117–147.
Wyrtki, K., J. B. Burks, R. C. Latham, and W. C. Patzert, 1967: Oceanographic observations during 1965–67 in the Hawaiian Archipelago. Univ. Hawaii, Tech. Rep. HIG-67-15, 150 pp.
Wyrtki, K., 1967: Oceanographic observations during the Line Islands expedition. Univ. Hawaii, Tech. Rep. HIG-67-17, 18 pp. + 7 figs.
Wyrtki, K., 1967: Water masses in the oceans and adjacent seas. *International Dictonary of Geophysics.* Pergamon Press, Tarrytown, NY, 1-11 + 8 figs.
Wyrtki, K., 1967: The spectrum of ocean turbulence over distances between 40 and 1000 kilometers. *German Hydrogr. J.*, **20** (4), 176–186.
Wyrtki, K., and K. Haberland, 1968: On the redistribution of heat in the North Pacific Ocean. *J. Oceanogr. Soc. Japan*, **24** (5), 220–233.
Wyrtki, K., V. Graefe, and W. Patzert, 1969: Current observations in the Hawaiian Archipelago. Univ. Hawaii, Tech. Rep. HIG-69–15, 97 pp.
Wyrtki, K., 1969: The duration of temperature anomalies in the North Pacific Ocean. *Bull. Japanese Soc. Fish. Oceanogr.* (Prof. Uda's commemorative papers), Univ. Hawaii HIG Contrib. No. 238, 81–86.
Wyrtki, K., 1970: Flights with airborne radiation thermometers in Hawaiian waters. Univ. Hawaii, Tech. Rep. HIG-70–5, 27 pp.
Patzert, W. C., K. Wyrtki, and H. J. Santamore, 1970: Current measurements in the Central North Pacific Ocean. Univ. Hawaii, Tech. Rept. HIG-70–31, 65 pp.
Wyrtki, K., 1971: *Oceanographic Atlas of the International Indian Ocean Expedition.* National Science Foundation Publication, OCE/NSF 86-00-001, Washington, DC, 531 pp.
Shaw, R. P., and K. Wyrtki, 1972: The shape of the warm surface layer in a subtropical gyre. *Studies in Physical Oceanography*, A. L. Gordon, Ed., New York, 179–194.
Wyrtki, K., 1973: Teleconnections in the equatorial Pacific Ocean. *Science*, **180**, 66–68.
Wyrtki, K., 1973: An equatorial jet in the Indian Ocean. *Science*, **181**, 262–264.
Wyrtki, K., 1973. Physical oceanography of the Indian Ocean: The Biology of the Indian Ocean, B. Zeitzschel, Ed., Springer-Verlag, Berlin, 18–36.
Wyrtki, K., 1974: Sea level and the seasonal fluctuations of the equatorial currents in the Western Pacific Ocean. *J. Phys. Oceanogr.*, **4**, 1, 91–103.
Wyrtki, K., 1974: On the deep circulation of the Red Sea. Colloques Internationaux du CNRS 215, Processus de Formation des eaux Océaniques Profondes, Univ. Hawaii HIG Controb. No. 481, 91–106.
Wyrtki, K., 1974: Equatorial currents in the Pacific 1950 to 1970 and their relations to the trade winds. *J. Phys. Oceanogr.*, **4** (3), 372–380.
Wyrtki, K., 1974: The dynamic topography of the Pacific Ocean and its fluctuations. Univ. Hawaii, Tech. Rep. HIG-74-5, 19 pp. + 37 figs.
Patzert, W., and K. Wyrtki, 1974: Anticyclonic flow around the Hawaiian Islands indicated by current meter data. *J. Phys. Oceanogr.*, **4** (4), 673-676.
Wyrtki, K., and G. Meyers, 1975: The trade wind field over the Pacific Ocean Part I: The mean field and the mean annual variation. Univ. Hawaii, Tech. Rep. HIG-75-1, 26 pp. + 38 figs.
Wyrtki, K., and G. Meyers, 1975: The trade wind field over the Pacific Ocean Part II: Bimonthly fields of wind stress: 1950 to 1972. Univ. Hawaii, Tech. Rep. HIG-75-2, 16 pp. + 132 figs.
Wyrtki, K., 1975: Fluctuations of the dynamic topography in the Pacific Ocean. *J. Phys. Oceanogr.*, **5** (3), 450–459.
Wyrtki, K., 1975: El Niño—the dynamic response of the equatorial Pacific Ocean to atmospheric forcing. *J. Phys. Oceanogr.*, **5** (4), 572–584.
Wyrtki, K., E. Stroup, W. Patzert, R. Williams, and W. Quinn, 1976: Predicting and observing El Niño. *Science*, **191** (4225), 343–346.

Wyrtki, K., L. Magaard, and J. Hager, 1976: Eddy energy in the oceans. *J. Geophys. Res.*, **81** (15), 2641–2646.

Wyrtki, K., and G. Meyers, 1976: The trade wind field over the Pacific Ocean. *J. Appl. Meteorol.*, **15** (7), 698–704.

Wyrtki, K., 1976: Climate fluctuations, ocean monitoring and buoys. Ocean Profiling Workshop, NOAA Data Buoy Office, Bay St. Louis, MS, 103–131.

Wyrtki, K., R. Bernstein, and W. White, 1976: NORPAX and the upper ocean. *Nav. Res. Rev.*, **29** (9), 1–18.

Wyrtki, K., G. Meyers, D. McLain, and W. Patzert, 1977: Variability of the thermal structure in the central equatorial Pacific Ocean. Univ. Hawaii, Tech. Rep. HIG-77-1, 75 pp.

Dotson, A., K. Wyrtki, L. Magaard, and G. Niemeyer, 1977: A simulation of the movements of fields of drifting buoys in the North Pacific Ocean. Univ. Hawaii, Tech. Rep. HIG-77-3, 59 pp.

Wyrtki, K., 1977: Advection in the Peru Current as observed by satellite. *J. Geophys. Res.*, **82** (27), 3939–3943.

Wyrtki, K., 1977: Sea level during the 1972 El Niño. *J. Phys. Oceanogr.*, **7** (6), 779–787.

Wyrtki, K., 1978: Monitoring the strength of equatorial currents from XBT sections and sea level. *J. Geophys. Res.*, **83** (C4), 1935–1940.

Wyrtki, K., 1978: Lateral oscillations of the Pacific Equatorial Countercurrent. *J. Phys. Oceanogr.*, **8** (3), 530–532.

Patzert, W., K. Wyrtki, T. Barnett, G. McNally, M. Sessions, B. Kilonsky, and D. Kirwan, 1978: Aircraft monitoring of ocean thermal structure and currents. *Nav. Res. Rev.*, **31** (9), 1–8.

Wyrtki, K., 1979: Sea level variations: Monitoring the breath of the Pacific. *EOS*, **60** (3), 25–27.

Wyrtki, K., 1979: Comments on the variability of the tropical ocean. *Dyn. Atmos. Oceans*, **3**, 209–212.

Wyrtki, K., 1979: The response of sea surface topography to the 1976 El Niño. *J. Phys. Oceanogr.*, **9** (6), 1223–1231.

Wyrtki, K., 1979: El Niño. *La Recherche*, **10**, 1212–1220.

Wyrtki, K., 1980: The Hawaii-to-Tahiti shuttle experiment. EDIS, Vol. ll, No. 6, 20–24.

Wyrtki, K., 1980: Hawaii-to-Tahiti shuttle experiment. *Mar. Weather Log,* **24** (5), 361–362.

Wyrtki, K., 1980: Scientific and operational requirements for monitoring the ocean– atmosphere environment by means of buoys. NOAA Data Buoy Office, NSTL Station, MS, F-821-1, 43 pp.

Wyrtki, K., and W. Leslie, 1980: The mean annual variation of sea level in the Pacific Ocean. Univ. Hawaii, Tech. Rep. HIG-80-5, 159 pp.

Wyrtki, K., 1980: Sea level during the NORPAX test shuttle experiment. Univ. Hawaii, Tech. Rep. HIG-80-6, 27 pp.

Wyrtki, K., E. Firing, D. Halpern, R. Knox, G. J. McNally, W. C. Patzert, E. D. Stroup, B. A. Taft, and R. Williams, 1981: The Hawaii-to-Tahiti shuttle experiment. *Science*, **211** (4484), 22–28.

Wyrtki, K., 1981: Comparison of four equatorial wind indices over the Pacific and El Niño outlook for 1981. *Proc. Fifth Ann. Climate Diagnostic Workshop* NOAA, Washington, DC, 211–218.

Stroup, E.D., K. Wyrtki, and B.J. Kilonsky, 1981: AXBT observations during the Hawaii-to-Tahiti shuttle experiments. Univ. Hawaii, Tech. Rep. HIG-81-1, 49 pp.

Wyrtki, K., 1981: An estimate of equatorial upwelling in the Pacific. *J. Phys. Oceanogr.,* **11** (9), 1205–1214.

Chaen, M., and K. Wyrtki, 1981: The 20°C isotherm and sea level in the western equatorial Pacific. *J. Oceanogr. Soc. Japan*, **37** (4), 198–200.

Wyrtki, K., 1982: The Southern Oscillation, ocean–atmosphere interaction and El Niño. *Mar. Tech. Soc. J.*, **6** (1), 3–10.

Wyrtki, K., 1982: Eddies in the Pacific North Equatorial Current. *J. Phys. Oceanogr.*, **12** (7), 746–749.

Wyrtki, K., and G. Eldin, 1982: Equatorial upwelling events in the Central Pacific. *J. Phys. Oceanogr.*, **12** (9), 984–988.

Wyrtki, K., and B. Kilonsky, 1982: Transequatorial water structure during the Hawaii-to-Tahiti shuttle experiment. Univ. Hawaii, Tech. Rep. HIG-82-5, 65 pp.

Wyrtki, K., and L. Uhrich, 1982: On the accuracy of heat storage computations. *J. Phys. Oceanogr.*, **12** (12), 1411–1416.

Wyrtki, K., 1983: An attempt to monitor the equatorial undercurrent. *J. Geophys. Res.*, **88** (C1), 775–777.
Wyrtki, K., 1983: Sea level in the equatorial Pacific in 1982. *Trop. Ocean Atmos.*, **16**, 6–7.
Firing, E., R. Lukas, J. Sadler, and K. Wyrtki, 1983: Equatorial undercurrent disappears during 1982–83 El Niño. *Science*, **222**, 1121–1123.
Wyrtki, K., 1984: A southward displacement of the subtropical gyre in the South Pacific during the 1982–83 El Niño. *Trop. Ocean-Atmos.*, **23**, 14–15.
Wyrtki, K., and B. Kilonsky, 1984: Mean water and current structure during the Hawaii-to-Tahiti shuttle experiment. *J. Phys. Oceanogr.*, **14** (2), 242–254.
Wyrtki, K., and J. Wenzel, 1984. Possible gyre–gyre interaction in the Pacific Ocean. *Nature*, **309** (5968), 538–540.
Wyrtki, K., and S. Nakahara, 1984: Monthly maps of sea level anomalies in the Pacific, 1975–1981. Univ. Hawaii, Tech. Rep. HIG 84-3, 8 pp. + maps.
Wyrtki, K., 1984: The slope of sea level along the equator during the 1982–83 El Niño. *J. Geophys. Res. Oceans*, **89** (C6), 10419–10424.
Lukas R., S. P. Hayes, and K. Wyrtki, 1984: Equatorial sea level response during the 1982–1983 El Niño. *J. Geophys. Res. Oceans,* **89** (C6), 10425–10430.
Wyrtki, K., 1985: Sea level fluctuations in the Pacific during the 1982–83 El Niño. *J. Geophys. Res. Lett.*, **12** (3), 125–128.
Wyrtki, K., 1985. Water displacements in the Pacific and the genesis of El Niño cycles, *J. Geophys. Res.Oceans*, **90**, 7129–7132.
Rebert, J. D., J. R. Donguy, G. Eldin, and K. Wyrtki, 1985: Relations between sea level, thermocline depth, heat content and dynamic height in the tropical Pacific Ocean. *J. Geophys. Res.*, **90** (C6), 11719–11725.
Wyrtki, K., 1985: Sea level data by satellite. *EOS*, **66** (32), 578.
Wyrtki, K., 1985: Monthly maps of sea level in the Pacific during the El Niño of 1982 and 1983. In: Time Series of Ocean Measurements, Vol. 2, IOC Tech. Series, 30, 43–54.
Wyrtki, K., 1985: Water displacements during 1982–1983 and the genesis of El Niño and the Southern Oscillation. In: International Conference on the TOGA Scientific Programme, UNESCO, World Climate Research Programme Series 4, WMO/TD 65, 1111–1110.
Wyrtki, K., 1985: Pacific-wide sea level fluctuations during the 1982–1983 El Niño. In: El Niño in the Galapagos Islands: The 1982–1983 Event. G. Robinson and E. del Pino, Eds., Charles Darwin Foundation for the Galapagos Islands, Quito, Ecuador, 29–48.
Donguy, J. R., G. Eldin, and K. Wyrtki, 1986: Sea level and dynamic topography in the Western Pacific during 1982–1983 El Niño. *Trop. Ocean Atmos.*, **36**, 1–3.
Bongers, T., and K. Wyrtki, 1987: Sea level at Tahiti—a minimum of variability. *J. Phys. Oceanogr.*, **17** (1), 164–168.
Wyrtki, K., 1987: Large-scale aspects of El Niño, in: Further progress in equatorial oceanography, E. Katz and J. Witte, Eds. (TOGA Workshop report). Nova Univ. Press, Ft. Lauderdale, FL, 259–262.
Wyrtki, K., 1987: Indices of equatorial currents in the Central Pacific. *Trop. Ocean Atmos.*, **38**, 3–5.
Wyrtki, K., 1987: Indonesian through flow and the associated pressure gradient. *J. Geophys. Res. Oceans,* **92** (C12), 12941–12946.
Wyrtki, K., 1987: Comparing GEOSAT altimetry and sea level. *EOS* , **68** (35), 731.
Wyrtki, K., K. Constantine, B. J. Kilonsky, G. Mitchum, B. Miyamoto, T. Murphy, S. Nakahara, and P. Caldwell, 1988: The Pacific Island Sea Level Network. Univ. Hawaii, JIMAR Contrib. No. 88-0137, Data Rep. 002, 71 pp.
Wyrtki, K., B. Kilonsky, and S. Nakahara, 1988: The IGOSS sea level pilot project in the Pacific. Univ. Hawaii, JIMAR Contrib. No. 88-0150, Data Rep. 003, 59 pp.
McPhaden, M., H. P. Freitag, S. Hayes, B. Taft, Z. Chen, and K. Wyrtki, 1988: The response of the equatorial Pacific Ocean to a westerly wind burst in May 1986. *J. Geophys. Res.*, **93** (C9), 10589–10603.
Mitchum, G. T., and K. Wyrtki, 1988: Overview of Pacific sea level variability. *Mar. Geod.*, **12**, 235–245.

Roach, D., G. Mitchum, and K. Wyrtki, 1989: Length scales of interannual sea level variability along the Pacific margin. *J. Phys. Oceanogr.*, **19** (1), 122–128.

Wyrtki, K., 1989: Sea level rise—the facts and the future. *Pac. Sci.*, **44** (1), 1–16.

Caldwell, C., K. Wyrtki, and S. Nakahara, 1989: TOGA Sea Level Center: Data from the Pacific. Univ. Hawaii, JIMAR Contrib. No. 89-0303, Data Rep. 6, 34 pp.

Wyrtki, K., 1989: Some thoughts about the West Pacific Warm Pool, in: J. Picaut et al., Eds., Proc. of Western Pacific International Meeting and Workshop on TOGA COARE. 99–109.

Wyrtki, K., and G. Mitchum, 1990: Interannual differences of Geosat altimeter heights and sea level: The importance of a datum. *J. Geophys. Res.* , **95** (C3), 2969–2975.

Wyrtki, K., 1993: Global sea level rise. Proc. Circum-Pacific Int. Symp. Earth Environment, National Fisheries Univ. Pusan, Pusan. D. Kim and Y. Kim, Eds., 215–226.

von Storch, H., J. Sündermann, and L. Magaard, 1999: Interview with Klavs Wyrtki, 25 Febuary 1999. GKSS Forschungszentrum Geestchacht, 42pp.

Index

Abyssal circulation, 20
Acoustic Doppler Current Profiler (ADCP), 54–55, 56–57, 58f, 69
Acoustic thermometry of ocean climate (ATOC), 131–33, 134
Acoustic tomography. *See* Ocean acoustic tomography
Acrivos, A., 46
Adamec, David, 110
ADCP. *See* Acoustic Doppler Current Profiler
Agnew, Spiro, 108
Agulhas Current, 2, 3
Agulhas Retroflection, 6
Agulhas Rings, 3, 6
Air Force, U.S., 186
Air-sea interaction, 53–56, 68–69
ALACE. *See* Autonomous Lagrangian Circulation Explorer
Alexander Agassiz (R/V), 123
Allen, John, 110
Altimetric measurements, 125, 133, 185–86
Amatek-Straza, 54
American Meteorological Society (AMS), 115
Amy Chouest (ship), 130
Anderson, David, 34, 110, 112
Anderson, G. C., 170, 175
Anderson, S. P., 69
Antarctica, 128, 166
Antarctic Bottom Water, 9
Antarctic Circumpolar Current, 3, 175
Antarctic Circumpolar front, 131
Antarctic Intermediate Water, 37, 171, 172
Anticyclonic gyres, 167–68, 173, 174
Antilles Current, 168
Antisubmarine warfare (ASW), 125–26
AOML. *See* Atlantic Oceanographic Marine Laboratory
Arabian Sea, 71
Arakawa, Akia, 32
Arctic Ocean, 4, 134, 166
Argo, 52, 62, 63, 70, 133, 134, 199
Aries (ketch), 47, 50
Aristotle, 2, 10
Atlantic-Indian Ocean exchange, 171
Atlantic Ocean, 11, 18, 95, 158, 172, 175. *See also* North Atlantic Ocean; South Atlantic Ocean
  acoustic tomography and, 134
  data collection in, 166
  dynamic calculations in, 8
  geopotential anomaly and, 172, 173, 174
  surface currents of, 3
  surface heat fluxes in, 72
  water properties, 4, 5, 6
Atlantic Oceanographic Marine Laboratory (AOML), 85, 86
ATLAS. *See* Autonomous Temperature Line Acquisition System
ATOC. *See* Acoustic thermometry of ocean climate
Autonomous Lagrangian Circulation Explorer (ALACE), 59–62, 185

Autonomous subsurface floats, 56, 57–59, 63
Autonomous Temperature Line Acquisition System (ATLAS), 54, 88, 90–91

Baker, D. James, 46, 104, 112, 183, 190
Baltic Sea, 213, 230–31
Barcilon, Victor, 103
Barnett, Tim, 54
Batchelor, George, 16
Battisti, David, 159
Behrens, Henry, 103
Bender, Peter, 185, 186
Berlin, Isaiah, 154
Bermuda, 25–26, 47, 48, 106, 107, 122, 123, 125, 128, 130, 169
Bermuda triangle, 25, 26
Bernard, Eddie, 91
Bernstein, Buzz, 54
Bezdeck, Hugo, 123
Bien, George, 166
Bierly, Gene, 109
Bigelow Laboratory, 102, 109
Birdsall, T., 122, 123, 128
Bjerknes, Jacob, 6, 80–81, 82, 95, 159, 210, 214, 217–18
Blandford, Robert, 102, 103
Bleck, R., 40
Blumberg, A. F., 40
Boening, C. W., 192
Bönecke, G., 205, 206, 207, 208
Booker, John, 16
Boudra, D., 40
Bowles, Ann, 130, 131
Boyd, John, 110
Bradley, Frank, 69
Brainard, Edward, 48
Brazil Basin Experiment, 192
Brazil Current, 173
Brekhovskikh, Leonid, 120–22, 126–28
Bretherton, Francis, 41, 50, 51, 57, 62, 83, 103, 106, 189, 190, 191
Bretherton Committee, 190
Brewer, P. M., 191
Briscoe, Mel, 56
Broecker, Wally S., 40, 166, 190, 191
Brogmus, Willi, 206
Brown, Neil, 208
Bryan, Frank, 38, 41
Bryan, Kirk, 158, 183
Buchan, A., 5
Buchanan, J. Y., 3
Bumble Bee buoys, 54
Buoy Group, 48–50, 52, 53–54, 56, 60, 62, 63
Buoy Project, 48, 55
Buoys
  Bumble Bee, 54
  CASID, 56
  Monster, 54
  PIRATA, 72
Burkov, V. A., 174
Bush, George H. W., 108

CAGE experiment, 83, 189, 197–98
California Current, 87
Cambridge University, 16, 34, 105, 124
Cane, Mark A., 107, 108, 158, 159
Cape Hatteras, 6, 168
Carbon-14, 18, 166
Carbon dioxide, 74, 75, 183, 191
Carnegie expedition, 174
Carpenter, T. H., 84
Carpenter, W. B., 10
Carrier, George F., 10, 102
Carritt, Dayton, 104, 109
Cartesian Diver, 61
CASID buoys, 56
CCCO. *See* Committee on Climate Change and the Ocean
Chain (R/V), 102, 107
Challenger expedition, 4, 155, 171
Challenger Reports, 5, 11
Charnell, R. L., 103
Charney, Jule G., 10, 30, 104, 105, 107, 145, 156, 160
Charney Committee, 183
Charnock, Henry, 103
Chereskin, T. K., 55
Chervin, Robert M., 41
Chesapeake Bay, 168, 169
Chlorofluorocarbons, 185
Chou, S., 72
CINCWIO. *See* Cooperative Investigations of the North and Central Western Indian Ocean
Circumpolar Current, 3, 8, 170, 175
Circumpolar front, 131
Circumpolar Water, 169–70
Clark, J., 122
Climate, 131–33, 184, 218, 221, 227–28, 229. *See also* Ocean general circulation models
Clinton, William, 104
CLIVAR, 159, 192, 199

Clowes, A. J., 8
COADS. *See* Comprehensive Ocean-Atmosphere Dataset
COARE. *See* Coupled Ocean-Atmosphere Response Experiment
Coats, D. A., 168
Columbus, Christopher, 2
Committee on Climate Change and the Ocean (CCCO), 184, 189, 190
Commonwealth Scientific and Industrial Research Organisation (CSIRO), 68, 73, 74
Comprehensive Ocean-Atmosphere Dataset (COADS), 71
Compton, Arthur Holly, 113
Cooperative Investigations of the North and Central Western Indian Ocean (CINCWIO), 110
Copley, Nancy, 148, 150
Core layers, 5
Cornuelle, Bruce, 126
Cory Chouest (ship), 128, 130, 131
Coupled Ocean-Atmosphere Response Experiment (COARE), 69, 70–73, 74
Cox, Charles, 54, 58, 61
Cox, Michael, 32, 33f, 35–36, 37, 41, 158
Crease, James, 47, 191
Cresswell, George, 68
Critical layer absorption, 16
Croll, James, 10
Cromwell, Townsend, 79, 208
Crowley, Pat, 33
CSIRO. *See* Commonwealth Scientific and Industrial Research Organisation
Current meters, 19, 20–21, 49. *See also* Vector Averaging Current Meter; Vector Measuring Current Meter

Dashen, S. M., 123
Davis, Russ E., 23–24, 106, 110
Deacon, G. E. R., 174, 207, 208
Deacon, Margaret, 11
Decca Hi-Fix, 104
Deep-Sea Research, 170
Deep Water, 4, 9, 168, 170, 175
Defant, A., 5, 120, 140, 167, 169, 173–74, 175, 203, 216
Demonstration Experiment, 126, 127f
De Solla Price, Derek, 154
Dietrich, G., 167, 168, 169, 173, 205, 206, 213
Discovery Investigations, 6, 8
Dobson, Fred, 189
Doppler shift measurements. *See* Acoustic Doppler Current Profiler
Dorson, Don, 59
Drake Passage, 8, 35, 217
Drifting thermistor chains, 83, 85–88
Droughts, 74, 90
Dufour, Jim, 60
Duing, Walter, 108–9, 110, 157
Duncan, Peter, 217
Dushaw, B. D., 132, 134
Dynamic calculations, 6–9
Dzieciuch, Matthew, 130

East Australian Current, 68, 209
East Greenland Current, 167
ECCO Consortium. *See* Estimating the Circulation and Climate of the Ocean Consortium
Eddies, 22–23, 36, 37, 50, 68, 105, 216–17, 226
Edmond, J. M., 170, 175
Ekman, V. W., 4
Ekman drifts, 3
Ekman layer, 16, 144, 218
Ekman pumping, 143, 145, 147
Ekman spiral, 10
Ekman transport, 165–66
El Chichon, 83, 84
Eleuthera, 122, 123
Ellen B. Scripps (R/V), 123
Ellett, D., 167, 175
El Niño-Southern Oscillation (ENSO), 35, 42, 53, 54, 79–95, 110, 112–13, 154, 155
  1982–83, 83, 84, 85, 158
  1986-87, 90
  advances in studies of, 159–61, 162
  drifting thermistor chains in studies, 83, 85–88
  first successful prediction of, 90
  phenomenon, 63, 68, 80, 90, 92, 93, 94, 95, 112, 159
  trade winds and, 80, 81, 84, 158
  Wyrtki on, 213, 214, 215, 224–25, 229, 231
EPOCS. *See* Equatorial Pacific Ocean Climate Studies
Equalant Atlas, 174
Equatorial dynamics, 113–15
Equatorial Pacific Ocean Climate Studies (EPOCS), 82–83, 88, 93, 95, 113
Equatorial Theoretical Panel, 80, 110, 113, 115

Equatorial Undercurrent, 3, 79, 80, 82, 83, 110, 111
  discovery of, 156
  generation simulations of, 158
  thermocline model of, 149
Equatorial waves, 103–4, 107–8, 110
ERS-1 satellite, 193
Esbenson, Steve, 69
Estimating the Circulation and Climate of the Ocean (ECCO) Consortium, 133
Ewing, J., 120, 121, 122, 125
Expendable bathythermographs (XBTs), 51, 54, 85–86, 88, 133, 199

Fairall, Chris, 69
Fandry, Chris, 51
Fanning Island, 105, 112
Farmer, David, 133
FGGE. *See* First GARP Global Experiment
FGGE/INDEX/NORPAX Equatorial (FINE) Workshop, 80, 110–11
Fieux, Michelle, 108
Findlater, J., 110
Findlay, A. G., 3
Fine, Rana, 112
FINE Workshop. *See* FGGE/INDEX/NORPAX Equatorial Workshop
Firing, Eric, 105
First GARP Global Experiment (FGGE), 109, 110, 113, 183–84, 190
Fischer, Gunter, 32
Flatté, R., 123
Fleming, J. A., 174
Fleming, R. H., 140
Fleming, Rex, 113
Fletcher, Joe, 162
Floats, 46–48, 50
  autonomous subsurface, 56, 57–59, 63
  Hippocampus, 58–59, 62
  MARVOR, 62
  PROVOR, 62
  RAFOS, 59, 60f, 62, 185
  SOFAR, 19, 21, 25, 50–62, 106, 123, 169, 185
Florida Current, 7, 104, 168
FOCAL. *See* Programme Français Océan et Climat dans l'Atlantique Equatorial
Fofonoff, Nick, 19, 48, 106
Forbes, Andrew M. G., 128, 130
Fram expedition, 4
Franklin (R/V), 69
Fredkin, Ed, 103
Frost, Jack, 111
Fuglister, Fritz C., 5, 11
Fung, Inez, 107

Galapagos Islands, 82, 158
Gandin, L. S., 24, 51
GARP. *See* Global Atmospheric Research Program
GARP Atlantic Tropical Experiment (GATE), 107, 108, 159
Garrett, C., 123
GATE. *See* GARP Atlantic Tropical Experiment
Gauss, Karl, 24
Gauss-Markov estimation theory, 126
General Circulation Models (GCMs), 158
Gent, Peter R., 40, 115
Geophysical Fluid Dynamics Laboratory (GFDL), 34, 37, 39, 40, 41, 109, 156
Geophysical Fluid Dynamics (GFD) program, 16–17, 102, 103, 109, 113, 148, 150–51
Geopotential anomaly, 172–74
GEOSAT, 90
GEOSECS, 106, 166, 213, 217, 226
Geostrophy, 6–9, 10
German Hydrographic Institute, 206
Gill, Adrian, 34–35, 80, 105, 116, 157
Glantz, M., 225
Global Atmospheric Research Program (GARP), 109, 183, 184, 218
Global ocean observations, 45–64
Global Ocean Observing System (GOOS), 95
Global warming, 134. *See also* Greenhouse warming
Global Weather Experiment. *See* First GARP Global Experiment
Gonella, Jose, 83
Goody, Richard, 103
GOOS. *See* Global Ocean Observing System
Gordon, Arnold L., 173, 209, 221
Gore, Al, 132
Gould, John, 63
Gould, W. J., 46
GRACE, 199
Greenhouse warming, 38, 73–77
  economic effects of, 73–75
  surface heat flux validation of, 75–77
Greenland Sea Project, 126
Gulf Stream, 2, 6, 46, 105, 108, 145, 157, 167, 168, 169, 173, 175, 186

general circulation models of, 35
MODE and, 24, 26
ocean acoustic tomography and, 126
Gulf Stream (R/V), 55, 109, 111–12

Hall, Mike, 85, 113
Halpern, David, 53–54, 56, 88, 91, 116, 158
Hamilton, Gordon, 122, 128
Hamon, Bruce, 68, 208
Hansen, W., 206
Harvard University, 36, 101, 102, 103, 107, 156
Harvey, Bob, 111
Hasselmann, Klaus F., 51, 52, 191
Hawaii Institute of Geophysics, 209, 211, 223
Hawaii Ocean Mixing Experiment (HOME), 132
Hawaii Sea Level Center, 113
Hawaii-to-Tahiti Shuttle Experiment, 82
Hayes, Stan, 62, 86, 88, 90, 91, 159
Hays, Earl, 102
Heard Island Feasibility Test (HIFT), 128–31, 132
Heinmiller, Bob, 48, 107, 112
Helium-3, 166, 185
Helland-Hansen, B., 4, 6, 9, 173
Hendershott, Myrl, 80, 110
Hersey, Brackett, 101
Hibler, Bill, 40
Hidaka, K., 207
Hill, Wayne, 55
Hippocampus floats, 58–59, 62
Hisard, Phillipe, 110, 111
HMAS *Diamantina*, 128
Holland, William, 26, 37, 41, 102, 145, 192
Holo Holo (cruise ship), 111–12
HOME. *See* Hawaii Ocean Mixing Experiment
Howard, Louis, 102
Huang, R. K., 149
Huppert, Herbert, 34
Hydrographic stations, 4, 6, 17, 18

Ibn Khurradadhbih, Abu'l Qasim, 2
ICSU. *See* International Council of Scientific Unions
IDOE. *See* International Decade of Ocean Exploration
IGY. *See* International Geophysical Year
IMET moorings, 69, 71, 72, 77
Imperial College, 103
INDEX, 108–10, 113
Indian Ocean, 69, 95, 105, 171, 188, 197
data collection in, 166
general circulation models of, 35, 37
geopotential anomaly and, 172, 174
INDEX and, 108–10, 113
surface currents of, 2, 3
surface heat fluxes in, 75–76
Indian Ocean Atlas, 210, 211–13, 230, 231
Inertia-gravity waves, 116
Ingersoll, Andy, 102
In situ measurements, 45, 105, 115
of El Niño, 85, 87, 91, 93
in ocean acoustic tomography, 133
of temperature, 4
in WOCE, 193, 194, 195
Institute for Advanced Study (IAS), 30
Institut für Meereskunde, 203–4
Intergovernmental Oceanographic Commission (IOC), 92, 108, 184
Intermediate Water, 37, 133, 171–72
International Council of Scientific Unions (ICSU), 92
International Decade of Ocean Exploration (IDOE), 15, 20, 53, 105, 106, 115, 156, 188, 213, 217
International Geophysical Year (IGY), 11, 187
International Ice Patrol, 8–9
International Indian Ocean Expedition, 208, 213
International Panel on Climate Change (IPCC), 39
International TOGA Scientific Steering Group, 91
Inverse theory, 123, 124, 176, 188
Inverted echo sounders, 21
IOC. *See* Intergovernmental Oceanographic Commission
Iorga, M. C., 171
IPCC. *See* International Panel on Climate Change
Irminger Sea, 167, 175
Isaacs, John, 54, 64
Iselin, Columbus, 9, 103, 140, 167, 168, 169, 175
Isopycnals, 40–41, 171
Ivers, W. D., 168, 171

Jacobs, Stan, 103
Japan Marine Earth Science and Technology Agency (JAMSTEC), 95
Jeffreys, H., 182
JIMAR. *See* Joint Institute for Marine and Atmospheric Research
JISAO. *See* Joint Institute for the Study of the Atmosphere and Oceans
Johns Hopkins University, 17

Johnson, M. W., 140
Joint Institute for Marine and Atmospheric Research (JIMAR), 111, 112, 115
Joint Institute for the Study of the Atmosphere and Oceans (JISAO), 115
Joseph, J., 206
Journal of Geophysical Research, 160
Journal of Marine Research, 108
Journal of Physical Oceanography, 115, 125
Journal of the Acoustical Society of America, 131
Joyce, Terry M., 55, 61
Jung, G. H., 11

Ka'imimoana (ship), 93
Karweit, Mike, 21, 24
Katz, Eli, 107, 113, 115
Kauai, Hawaii, 132, 133, 134
Keeling, C. D., 191
Kelvin, Lord, 79, 84, 95
Kelvin waves, 81, 83, 90, 94, 107, 116, 157, 158, 214
Kennedy, Ted, 105
Killworth, P., 148–49
Kimura, R., 191
Kirwan, Denny, 105
Knauss, John, 108, 210
Knox, Robert, 80, 82, 92, 116, 157
Knox Report, 92
Koenuma, K., 173, 174
Kraus, Eric, 83, 103
Kronengold, M., 122
Krümmel, O., 120
Kuperman, W., 131
Kuroshio Current, 6, 8, 171

Lacombe, H., 174
Lagrangian coordinates, 40
Lagrangian statistics, 52
Langmuir circulation, 56
La Niña, 90, 94, 159, 224
Large, Bill, 56, 87
Lautenbacher, Conrad, 105
Lawrence Livermore Laboratory, 33
Lawson-Dick, Oliver, 11
Lee, A., 167, 175
Leetmaa, Ants, 157, 159
Lefebvre, M. P., 191
Legeckis, Richard, 157
Level of no motion, 18, 172, 188
Levitus, S., 171
Lewis, Larry J., 37, 38
Lighthill, M. J., 107, 157
Lilly, Douglas, 31–32
Lin, C. C., 16
Lindzen, Richard S., 102, 104, 108
Longuet-Higgins, Michael, 103, 128
Loran-A radio-navigation system, 6
Lorenz, E. N., 182
Lozier, M. S., 171
Lukas, Roger, 68
Luyten, Jim, 105, 110, 112, 139, 145, 146, 148
Lynch, Jim, 126, 132
Lynn, R. J., 171

Madden-Julian Oscillations, 68, 75, 94
Magaard, Lorenz, 232
Maksimov, I. V., 174
Malkus, Joanne, 102
Malkus, Wilhelm, 102
Manabe, Syukuru, 31, 37, 38, 39
Mantyla, A. W., 171
Marine Mammal Research Program (MMRP), 132
Martec group, 58
MARVOR floats, 62
Massachusetts Institute of Technology (MIT), 16, 103, 105, 107, 108, 112, 123, 156
Masuzawa, J., 171
Matsuno, Taroh, 103, 157
Maximum-minimum thermometers, 4
Max Planck Institute, 40
McCallister, T., 170
McCreary, Julian (Jay), 80, 110, 111, 112, 115, 154
McDonald, A. M., 170
McDougall, T. J., 171
McEwen, G. F., 173, 174
McNally, Gerald, 55
McPhaden, Michael J., 46, 62, 111, 115
McWilliams, Jim, 26, 40, 52, 87, 190
Media, 224
Mediterranean Sea, 5, 101–2, 133, 167, 168, 175
Mellor, G. L., 40
Mercury thermometers, 4
Merle, Jacques, 111, 154
Merrifield, Mark, 113
Merz, A., 5
Mesoscale, 36, 37, 50, 120, 124, 126, 128, 133
Meteor Atlas, 167, 173
Meteor expedition, 5, 8, 167

Metzger, K., 123, 126, 130
Middle North Atlantic Deep Water, 168, 170, 175
Mid-Ocean Dynamics Experiment (MODE), 15–27, 37, 50–52, 53, 56, 105–7, 115, 120, 126, 185, 188, 213, 226
 Field Program, 21, 24, 25–26
 influence on WOCE, 182–83
 MODE-0 phase, 24, 50, 51–52
 preparation for, 20–25
 revolutionary design of, 51
 roots of, 17–20
 Scientific Council, 24
 Theoretical Panel, 21, 23, 24
"Mid-Ocean Madness" (Richardson), 105
Mikhalevsky, P. N., 133
Milankovitch cycles, 161
Milburn, Hugh, 53–54, 64, 82, 88
MILE. *See* Mixed Layer Experiment
MIMI transmission, 122, 123
Mintz, Yale, 32, 36
Mitchum, Gary, 113
Mixed Layer Experiment (MILE), 56
Moana Wave (R/V), 69
MODE. *See* Mid-Ocean Dynamics Experiment
Modular Ocean Model (MOM), 37
Mohn, H., 6
Monsoons, 3, 75, 95, 110, 197
Monster Buoys, 54
Montgomery, R. B., 5, 140, 167, 168, 171, 173, 174, 175
Mooers, Chris, 115
Mooney, K., 113, 168
Moore, Dennis, 24, 25, 80, 154, 157, 158
Moored thermistor chains, 86
Moore's Law, 32, 187
Moorings, 47–49, 52, 54, 126
Morawitz, W. M. L., 126
Morison, S. E., 2
Moura, Antonio, 107
Müller, Peter, 232
Munk, Walter H., 10, 29, 32, 45, 104, 167, 168, 175, 185, 190
Murray, J., 5, 11

NAGA expedition, 101
NAGA Report, 207, 231
Namias, Jerome, 10, 54, 81, 214, 217–18
Nansen, F., 4, 173
NASA, 77, 90, 109, 185–86, 194, 195
National Academy of Sciences (NAS), 35, 39, 106, 190
National Atmospheric and Oceanic Agency (NOAA), 53, 54, 104, 108, 109, 115, 185
 El Niño studies and, 82, 84, 85, 90, 93, 95, 162
 MODE and, 15, 25
 ocean acoustic tomography and, 130
National Center for Atmospheric Research (NCAR), 22, 24, 26, 32, 33, 41, 50, 56, 83, 105, 106
National Data Buoy Center, 109
National Data Buoy Project, 107
National Institute of Oceanography (NIO), British, 46, 106
National Marine Fisheries Service, 130
National Meteorological Center, 90
National Research Council, 92, 183, 185, 190
National Science Foundation (NSF), 16, 54, 87, 105, 109, 112, 184, 211, 213, 226
 MODE and, 15, 20
 NORPAX and, 81
 WOCE and, 60
Nature, 125
Navy, U.S., 19, 90, 93, 121, 125–26, 128. *See also* Office of Naval Research
Navy, U.S.S.R., 121
NCAR. *See* National Center for Atmospheric Research
Needler, G. T., 142
Neelin, David, 159, 160
Neumann, G., 206
Neumann, John von, 30
Neutral surfaces, 171
Nicholls, Neville, 68
Niiler, Pearn (Peter), 56, 59, 86, 87, 104, 105, 108, 109
NOAA. *See* National Atmospheric and Oceanic Agency
NORPAX. *See* North Pacific Experiment
North Atlantic Current, 62
North Atlantic Deep Water, 4, 9
North Atlantic Intermediate Water, 133
North Atlantic Ocean, 18, 71, 83, 171, 175, 189, 192, 196
 acoustic tomography and, 128
 bottom water of, 167
 general circulation models of, 35, 36, 40, 41
 subpolar gyre of, 11
 water properties, 6
North Atlantic Study (NAS), 166

North Indian Ocean, 3
North Pacific Acoustic Laboratory (NPAL), 133–34
North Pacific Experiment (NORPAX), 54, 55, 81–83, 93, 95, 106, 111, 113
  establishment of, 81
  Wyrtki on, 210, 213, 215, 217, 224, 226
North Pacific Intermediate Water, 171–72
North Pacific Ocean, 54, 85, 128, 168, 174, 175, 192, 211
Norwegian Sea, 167, 175
Nova University, 55, 104, 105, 108, 109–10, 111, 112
Nowlin, Worth, 105, 190, 193
NPAL. *See* North Pacific Acoustic Laboratory

Obasi, G. O. P., 93
O'Brien, Jim, 32, 80, 105, 110, 111, 115
Ocean acoustic tomography, 119–36
Ocean general circulation models (OGCMs), 29–42, 159
Ocean Modelling, 148
Ocean Observing System Development Panel (OOSDP), 91
Ocean Observing System for Climate (OCEANOBS), 134
Oceanography Society, 128
Ocean Researcher (R/V), 91
Ocean Storms, 56, 57, 58f
Ocean Studies Board (OSB), 190
Office of Global Programs, 113
Office of Naval Research (ONR), 54, 57, 101, 105, 109, 111, 123, 211
  MODE and, 15
  NORPAX and, 81
  ocean acoustic tomography and, 125
  WOCE and, 184, 192
Okhotsk Sea, 172
Ollitrault, Michelle, 62
OOSDP. *See* Ocean Observing System Development Panel
ORION, 134–35

Pacanowski, Ron, 37, 156
Pacheco, Claude, 58, 60
Pacific Decadal Oscillation, 54
Pacific Equatorial Ocean Dynamics (PEQUOD), 105, 112–13
Pacific Marine Environmental Laboratory (PMEL), 53, 56, 62, 63, 64, 82, 87, 88, 91, 111, 115
Pacific Ocean, 18, 68, 79, 80, 81, 84, 85, 86–87, 88, 113, 157, 158, 159, 171, 213, 214. *See also* El Niño; North Pacific Ocean; South Pacific Ocean; Tropical Ocean Global Atmosphere system
  acoustic tomography and, 131–32, 134
  air-sea interaction in, 53–54
  data collection in, 166
  geopotential anomaly and, 172, 173, 174
Paleoclimates, 161
Parr, A. E., 167, 171, 173
Paulson, Clayton, 69
Pawlowicz, R., 126
Pedlosky, Joseph, 102
PEQUOD. *See* Pacific Equatorial Ocean Dynamics
Perth, Australia, 73–75, 128, 130
Peru current, 230
Peruvian anchovy fishery, 81
Philander, George, 35, 104, 107, 108, 109, 110
Phillips, Norm, 105
Phillips, Owen, 16
Philosophical Transactions of the Royal Society, 111
Picaut, Joel, 80, 88, 111
Pierce, S. D., 55
Pillsbury, J. E., 7
Pinkel, R., 54
Pioneer Seamount, 132
PIRATA buoys, 72
Pittenger, Richard, 128
PMEL. *See* Pacific Marine Environmental Laboratory
Pollak, M. J., 167, 168, 175
POLYMODE, 27, 51, 183, 226
Ponce de Leon, 2
Porter, M., 123
Potential temperature, 4
Potter, John, 132
Price, J. F., 60
Price, Jim, 52
Princeton University, 34, 36, 37, 38, 40, 41, 84, 109, 112
Programme Français Océan et Climat dans l'Atlantique Equatorial (FOCAL), 83, 85, 87
PROVOR floats, 62

QuikSCAT, 72

Rainfall, 74, 75
Ramage, Colin, 110
Rasmusson, E. M., 84
Rattray, Maurice, 103, 104, 106
RD Instruments (RDI), 54, 55, 56, 57
Redi, M. H., 40
Regier, Lloyd, 54
Reichelderfer, Francis, 31
Reid, Joseph L., 8, 210
Rennell, J., 3
Revelle, Roger, 183, 207–8
Reversing thermometers, 4
Rhines, Peter, 37, 41, 143–45, 147, 149, 168
Richardson, L. F., 29–30
Richardson, P. L., 168, 169
Richardson, William, 47–48, 49, 55, 64, 104, 105, 109
Richter, Frank, 145
Riley, G. A., 11
Robinson, Allan, 36, 50, 103, 104, 105, 106, 141, 142, 156
Rocheford, David, 208, 213
Roemmich, Dean, 63, 170, 188n5
Roll, H., 206
Rooth, Claes, 34, 38, 39, 40, 102, 160
Rossby, C. G., 140, 168–69, 173, 175
Rossby, Tom, 19, 25, 50, 52, 53f, 59, 103
Rossby waves, 90, 103, 157
Rothstein, Lew, 113–15
Rotschi, H., 174
Rowlands, Philip, 108
Royal Navy, British, 46
Royal Navy, Thai, 101

Saito, Y., 174
Salinity, 4, 5, 7, 61
Salinity, Temperature, Depth. *See* STD
Salmon, R., 149
Saluda (ship), 120
Samelson, R., 149
Samudera (research vessel), 206
Sandström, J. W., 6
Sarachik, Ed, 107–8, 110, 154
Sargasso Sea, 22
Sarkisyan, Artem, 32, 35, 191
Sarmiento, Jorge, 40
Satellites, 120, 125, 133, 219–20, 226–27
  in El Niño studies, 84
  ERS-1, 193
  Sputnik, 153, 154, 156, 159
  in surface heat flux studies, 71–72
SAVE. *See* South Atlantic Ventilation Experiment
Scatterometers, 71–72
SCAVE. *See* Sound Channel Axis Experiment
Schäfer, Benny, 210
Schmitz, Bill, 51, 52, 64, 104–5
Schopf, Paul, 159
Schott, Fritz, 8, 57, 110, 112, 221
Schott, G., 3
Schulman, Elliott, 105
Science, 113
Scientific Committee on Oceanic Research (SCOR), 110, 112, 133, 184
Scorpio expedition, 166
Scripps Institution of Oceanography (SIO), 54, 56, 57, 59, 64, 79, 80, 82, 101, 102, 104, 105, 121
  ocean acoustic tomography and, 123, 132
  WOCE and, 60, 61
  Wyrtki on, 209, 210–11, 226
Sea Bird, 61
SEASAT spacecrafts, 185–86
Seasonal Response of the Equatorial Atlantic (SEQUAL), 83, 113
Semtner, A. J. (Bert), 36–37, 41, 192
SEQUAL. *See* Seasonal Response of the Equatorial Atlantic
Service Argos, 86, 88
Sextants, 153
Shadow Zone, 147, 149
Sherman, Jeff, 60
Sigma ocean circulation models, 40–41
Slocum (glider), 60
Smagorinsky, Joseph, 30, 31, 32, 34, 36, 37, 109, 156
Smith, E. H., 173
Snyder, Russ, 104
SOFAR channel, 46, 50, 120
SOFAR floats, 19, 21, 25, 50, 51, 52, 53f, 56, 59, 60f, 62, 106, 123, 169, 185
SOFAR overture, 135
SOLO. *See* Sounding Oceanographic Lagrangian Observer
Solomon, H., 40
Somali Current, 157
SOSUS array, 125–26
Sound Channel Axis Experiment (SCAVE), 122
Sounding Oceanographic Lagrangian Observer (SOLO), 61–62
Southampton Oceanography Centre (SOC), 71, 72

South Atlantic Ocean, 4, 5, 11, 170, 172, 192
South Atlantic Ventilation Experiment (SAVE), 166
Southern Ocean, 4, 6, 8, 35, 60, 213
Southern Oscillation. *See* El Niño-Southern Oscillation phenomenon
South Indian Ocean, 168, 172
South Pacific Ocean, 8, 62, 128, 168, 172, 193
Soviet Union, 154–55
Speed logs, 54
Spelman, Mike, 38
Spiesberger, J. L., 126, 128
Spindel, R., 123, 130, 132
Sputnik (satellite), 153, 154, 156, 159
SST. *See* Sea Surface Temperature
Starr, Victor, 182
STD, 19–20, 21, 25, 51, 52, 104
Steele, J., 190
Steinberg, C., 122
Stern, Melvin, 102, 148
Stewart, Robert, 189, 197–198
Stommel, Henry (Hank), 8, 10, 11, 16, 17, 18, 20, 29, 32, 38, 46–47, 50, 60, 102–3, 104, 105, 106–7, 108–9, 110, 113, 115, 139, 141, 144–45, 146–49, 156, 157, 160, 166, 168, 174, 183, 197–98n9, 218, 219, 220, 226
Stone, Peter, 104
Storch, Hans von, 232
Storm Transfer and Response Experiment (STREX), 56, 85, 87
Straits of Florida, 7, 122
Straits of Gibraltar, 5, 167, 168
STREX. *See* Storm Transfer and Response Experiment
Suarez, Max, 159
Subarctic gyres, 3, 8, 173
Subduction Experiment, 192
Sumi, Akimasa, 88
Sündermann, Jürgen, 232
Surface drifters, 56–57, 59
Surface heat fluxes, 67–77
  greenhouse model validation with, 75–77
  measuring and checking, 69–70
  mismatch problem in study, 70
  TOGA-COARE accuracy transfer and, 70–73
Surface Velocity Program, 62
Sutton, P. J., 126
Sverdrup, Harrold U., 7, 9–10, 11, 29, 120, 140, 144, 156, 167, 169, 174, 175, 208, 216
Sverdrup relation, 145, 146
Sverdrup transport, 144, 166, 167, 213
Swallow, John, 17, 19, 20, 21, 46–47, 50, 62, 63, 105, 110, 169, 174
Swamp models, 184
Synoptic scales, 35

Taft, Bruce, 53–54, 86, 87, 110, 115
Takano, K., 35, 36, 37
Takeuchi, Kensuke, 88
Talley, L. D., 170, 172
TAO Array. *See* Tropical Atmosphere Ocean array
TAO Implementation Panel (TIP), 91, 92
Tasman Sea, 131, 213
Tellus, 141
Temperature, 4, 5, 6, 7, 61. *See also* Sea Surface Temperature; STD
Thayer, Mary, 17
Thermistor chains
  drifting, 83, 85–88
  moored, 86
Thermocline theory, 139–51
  layer models of, 143–49
  modern (1959–80), 141–43
Thermohaline circulation, 38, 230
THETIS-2, 133
Thompson, R. J., 181, 184
TOGA. *See* Tropical Ocean Global Atmosphere system
Toggweiler, Robbie, 40
Tomczak, M., 206
TOPEX, 214, 216, 220
TOPEX/POSEIDON, 133, 186, 193, 195
TOPEX Science Working Group, 196
Trade winds, 80, 81, 84, 158, 214
Transient Tracers in the Ocean (TTO), 166
TRANSPAC XBT program, 54
Tritium, 185
TRITON Array, 62–63, 95
Tropical Atmosphere Ocean (TAO) array, 54, 57, 62–63, 88–95, 159, 199, 227
Tropical Ocean Global Atmosphere (TOGA) system, 46, 52, 56, 62, 88–94, 113, 189, 199, 227. *See also* Tropical Atmosphere Ocean array
Tropic Heat, 113
Tsuchiya, M., 170, 171, 172
Tsunamis, 113
Turner, Stuart, 46

Uda, M., 174
University-National Oceanographic Laboratory System (UNOLS), 109, 111–12
University of California, Los Angeles (UCLA), 32, 33, 36, 37
University of Hawaii, 111
University of Michigan, 123
University of Rhode Island, 115
UNOLS. *See* University-National Oceanographic Laboratory System
Upper North Atlantic Deep Water, 168
USNS *Titan*, 93

VACM. *See* Vector Averaging Current Meter
Vallis, G., 149
Vector Averaging Current Meter (VACM), 50, 106
Vector Measuring Current Meter (VMCM), 55–56, 57, 58f
Vema (R/V), 128
Ventilated Thermocline, theory of, 139, 146
Veronis, George, 16, 39–40, 106, 108, 142, 157, 183
Verstraete, Jean Marc, 111
VMCM. *See* Vector Measuring Current Meter
Volunteer Observing Ships, 72–73

Walker, Sir Gilbert, 80
Warren, Bruce, 140, 166, 193
Water masses, 9–10
Water properties, 4–6
Water resources, 227–28
Wattenberg, H., 5
Weather Bureau, U.S., 30, 31–34, 35, 41
Webb, David, 68
Webb, Doug, 50, 52, 53f, 56, 57–59, 60–61
Webb Research, 61
Webster, F., 190
Webster, Ferris, 48
Webster, Peter, 68
Wecoma (R/V), 69
Weddell gyre, 3
Weidermann, H., 206
Welander, P., 142, 143, 149, 167, 168, 175
Weller, Robert, 56, 69
WEPOCS, 113
Western intensification, 166
Wexler, Harry, 31
Whales, 132
White, Warren, 54
Willebrand, Juergen, 34, 39, 40
Wimbush, Mark, 109
Witte, Jan, 109, 110, 112, 113
WMO. *See* World Meteorological Organization
WOCE. *See* World Ocean Circulation Experiment
Wollard, George, 211
Woodcock, Al, 102
Woods, J., 191
Woods Hole Oceanographic Institution (WHOI)
  Bretherton on, 16–17, 19
  Bryan on, 31
  Davis on, 46, 47, 48, 50, 51, 52, 53, 56, 62
  Godfrey on, 72, 73, 75, 76
  McPhaden on, 82
  Moore on, 101, 102, 103, 104, 106, 107, 108, 109, 112, 113
  Munk on, 120, 121, 123, 132
  Pedlosky on, 143, 148
  Warren on, 11
  Wyrtki on, 226
Wooster, W., 210
Worcester, P. F., 123, 126, 128, 133
World Climate Research Program (WCRP), 191
World Meteorological Organization (WMO), 92, 93
World Ocean, 35–36, 37, 41, 42, 167, 174
World Ocean Circulation Experiment (WOCE), 34, 56, 63, 73, 77, 133, 166, 181–200
  background science, 182–85
  CAGE experiment and, 83
  development of, 59–62
  Hydrographic Survey, 60
  modeling and theory, 187–88
  objective of, 59
  origins of, 182–88
  planning for, 190–95
  proposing and selling of, 189–90
World Weather Watch, 92
Worthington, Hal, 17, 18, 20
Worthington, L. V., 46–47, 167, 168, 169, 174, 175
Worzel, Joseph, 120
Wright, W. R., 167, 175
Wunsch, Carl, 50, 55, 80, 105, 108, 110, 116, 120, 123, 124–25, 157, 216

Wüst, Georg, 5, 7–8, 11, 167, 168, 169, 170, 173, 175, 204, 206, 208, 209, 212, 220, 229
Wyrtki, Klaus, 81, 83, 84, 110, 111, 113, 158, 174, 203–32
  on changing themes, 216–18
  on funding, 223–24
  on the future of science, 226–27
  on impediments to progress, 226
  Indian Ocean Atlas and, 210, 211–13
  on the media, 224
  on models, 218–19, 228–29
  on politics, 224
  on the role of colleagues, 225–26
  on the role of government, 220–21
  university studies of, 203–5

XBTs. *See* Expendable bathythermographs

Yale University, 107
Yanai waves, 107
Yentsch, Charlie, 104, 109
You, Y., 171, 172
Young, William, 143–44, 145, 147, 149, 168
Yu, Lisan, 72, 76

Zebiak, Steve, 159